Joseph Long

American wild-fowl shooting

Describing the haunts, habits, and methods of shooting wild fowl

Joseph Long

American wild-fowl shooting
Describing the haunts, habits, and methods of shooting wild fowl

ISBN/EAN: 9783337174439

Printed in Europe, USA, Canada, Australia, Japan

Cover: Foto ©berggeist007 / pixelio.de

More available books at **www.hansebooks.com**

AMERICAN
WILD-FOWL SHOOTING.

DESCRIBING THE

HAUNTS, HABITS, AND METHODS OF SHOOTING
WILD FOWL, PARTICULARLY THOSE OF THE
WESTERN STATES OF AMERICA.

WITH

INSTRUCTIONS CONCERNING GUNS, BLINDS, BOATS, AND
DECOYS; THE TRAINING OF WATER-
RETRIEVERS, ETC.

BY
JOSEPH W. LONG.

NEW YORK:
J. B. FORD AND COMPANY.

1874.

TO MY FRIEND

FRED KIMBLE,

of Peoria, Illinois,

A CRACK DUCK-SHOT AND AN HONEST MAN,

THIS BOOK

IS RESPECTFULLY DEDICATED

BY

Its Author.

CONTENTS.

		PAGE.
INTRODUCTION,	. . . .	13

CHAPTER I.
| GUNS, | . . . . | 19 |

CHAPTER II.
| BLINDS, | . . . . | 45 |

CHAPTER III.
| DECOYS, | . . . . | 55 |

CHAPTER IV.
| BOATS AND BOAT-BUILDING, | . . . . | 78 |

CHAPTER V.
| DOGS, | . . . . | 93 |

CHAPTER VI.
| CAMPING OUT, | . . . . | 101 |

CHAPTER VII.
| MISCELLANEOUS HINTS, | . . . . | 110 |

PAGE.

CHAPTER VIII.
MORNING MALLARD SHOOTING—FALL, ... 125

CHAPTER IX.
MID-DAY MALLARD SHOOTING—FALL, ... 149

CHAPTER X.
EVENING MALLARD SHOOTING—FALL, ... 164

CHAPTER XI.
CORN-FIELD MALLARD SHOOTING—WINTER, . . . 169

CHAPTER XII.
MALLARD SHOOTING AT HOLES IN THE ICE—WINTER, . . 173

CHAPTER XIII.
MALLARD SHOOTING IN THE TIMBER—SPRING, . . 177

CHAPTER XIV.
BLUE-WINGED TEAL, ... 182

CHAPTER XV.
GREEN-WINGED TEAL, ... 190

CHAPTER XVI.
PINTAIL DUCK, ... 195

CHAPTER XVII.
WOOD DUCK OR SUMMER DUCK, ... 201

PAGE.

CHAPTER XVIII.

AMERICAN WIDGEON, ... 206

CHAPTER XIX.

GADWALL DUCK, ... 211

CHAPTER XX.

SHOVELLER DUCK, ... 217

CHAPTER XXI.

DUSKY DUCK, ... 221

CHAPTER XXII.

TRUMPETER SWAN, ... 224

CHAPTER XXIII.

CANADA GOOSE, ... 232

CHAPTER XXIV.

WHITE-FRONTED GOOSE, ... 241

CHAPTER XXV.

CANVAS-BACK DUCK, ... 246

CHAPTER XXVI.

RED-HEADED DUCK, ... 265

CHAPTER XXVII.

SCAUP DUCK, ... 271

PAGE.

CHAPTER XXVIII.

RING-NECKED DUCK, ... 276

CHAPTER XXIX.

BUFFLE-HEADED DUCK, ... 279

CHAPTER XXX.

FINALE—CAVE CORVUM ! 283

INTRODUCTION.

In the following work 1 have endeavored to lay before the public, in as concise a form as possible, full and trustworthy explanations of the various practical methods of hunting wild fowl as followed in the inland portions of our country. No book that I am yet aware of, published in this country, has been devoted exclusively to the subject; though a few short, fragmentary articles, giving a general idea of the sport, are to be found in the many volumes of our sporting literature.

Wild-fowling as an art is but very little understood by the great majority of sportsmen. It is attended with too much fatigue and too many hardships to be followed by them as it should be in all weathers, in order to become thoroughly familiar with it. And even in the warmer months, nearly all of our richer and better-educated sportsmen, instead of making use of their own reasoning powers to find and secure their game, depend on hiring others more experienced and willing to work, and do not question the whys and wherefores of their movements, or those of the game. They are content simply to shoot ·when opportunities present themselves. The man who is hired does not consider it his particular duty to explain the various habits of the birds, nor the proper manner of taking advantage of a knowledge of them. Amongst professional hunters, for various reasons, thoughts of book-making are seldom entertained; and though there are many excellent writers, both amateur and professional, who understand it much better than myself, and are much better able to do it justice, the subject—one of unusual interest to American sportsmen—has been almost entirely neglected. Such being the case, it is not without a well-grounded hope that, in the absence of such a work as *might* be, my labors may not prove altogether useless.

The ornithological descriptions, by referring to which the novice may determine the specific characteristics of his game, 1 have borrowed from Audubon's "Biographical Ornithology"; farther than this 1 have abstained from copying from other authors, and contented myself with writing only that which 1 have learned and proved by experience. 1 have aimed to instruct rather than to amuse, leaving it to others more capable to discourse upon the poetry and romance of the sport, and retaining only the less enticing but more profitable philosophy and reason; and, as I am not accustomed to literary pursuits, I trust that anything I may have written that appears egotistical or self-assuming may be ascribed by my generous critics, sportsmen, to a visibly poor acquaintance with the graces of rhetoric and style.

Before treating of the various methods to be employed in the pursuit of wild fowl, 1 shall first proceed to separate them into two distinct classes, which I shall term respectively the *deep-water* and *shoal-water* varieties, taking the Canvas-back as the type of the one, and the Mallard of the other. The habits of the two varieties vary so greatly that many rules which may be employed successfully in the pursuit of the one, it will be readily seen, might not be applicable to the killing of the other.

The shoal-water varieties simply immerse the head and neck, but seldom or never entirely submerge the body when feeding, though having the power of doing so and swimming under water quite long distances when wounded and pursued. Their food, then, of course, must be different and their natural haunts separate from those of the deep-water varieties, which, as their name indicates, dive for their food. The bodies of the deep-water varieties are proportionally broader, both as compared with their length and depth; the legs set farther back, and the wings proportionally shorter than those of the other class; the tail feathers shorter, more stiff, and less inclined to " cock up," being naturally held horizontally or slightly drooping; the plumage is generally shorter, but the feathers are more close and densely filled with down; and as a rule they will be found more tenacious of life than the shoal-water ducks. In the deep-water varieties, with the canvas-back 1 shall treat of the red-head, blue-bill or broad-bill, tufted duck, and buffle-head or butter-ball.

In the shoal-water class are the mallard, sprig-tail or pintail, blue and green winged teal, wood or summer duck, gadwall or gray duck, widgeon, shoveller, and dusky or black duck (the last-named, though not properly a Western duck, being sometimes found associated with the mallard). To the shoal-water variety, with the ducks belong the wild goose and the swan, whose habits, with the manner of hunting them, I shall also consider.

It will be noticed by those conversant with Western wild-fowling that two or three species of ducks, with the mergansers found on Western waters, have been omitted in the above classification. These I have considered it a great waste of time and space to describe, as they are rarely pursued for sport, and are of such rank and fishy flavor as to be totally unfit for the table.

Though several ornithologists have made this same division of the duck family, as may be seen by referring to their scientific nomenclature, they have, however, denominated their subfamilies Fuligulinæ, or sea-ducks, and Anatinæ, fresh-water ducks; and in view of the fact that I am treating solely of *inland* wild-fowling, and yet find both classes in nearly equal abundance, I have deemed it more appropriate to their habits and general characteristics to classify them as I have at first done. I have dwelt at greatest length upon the pursuit of the mallard and canvas-back, as it is to the capture of these two varieties,

representative types of the two classes, that the
labors of sportsmen are chiefly directed. Many of
the methods therein described will be found applicable to the taking of the other varieties; but
where the different habits of the other varieties
occasion different plans of operation, they are
elsewhere duly and explicitly explained.

The various concurrent subjects of guns, boats,
decoys, etc., have been briefly yet compre-
hensively discussed under their proper headings.

CHAPTER I.

GUNS.

I SUPPOSE I must say a few words as to the comparative merits of breech and muzzle loaders; for, to my mind, notwithstanding I am willing to acknowledge the general superiority of the new invention, the muzzle-loader possesses several decided advantages which entitle it to merit, and, in certain cases, to preference in selecting the most suitable weapon.

The chief superiority of the breech-loader lies in its capability of being so quickly reloaded when in the field or boat, and this alone is a sufficient advantage to compensate for many otherwise serious objections; for, no matter how smart a man may be with a muzzle-loader, he will often lose many opportunities of shooting, through not being ready loaded, while to the patron of the breech-loader such occurrences are very rare, provided he has plenty of loaded cartridges handy. This one requirement, however, being often wanting, the advantage is not so conclusive. The providing or reloading of a sufficient number of metallic shells, or the transportation and room required for their paper substitutes, is the most serious drawback in their use; and here it is that the claim of the muzzle-loader is especially noticeable. This con-clusion was not formed hastily, but was brought to my mind very forcibly, though rather unwill-ingly, and most frequently whilst sitting up at night loading shells and listening to the snoring of my fellow-hunters, votaries of the muzzle-loader, who, having eaten their supper, washed their guns, and refilled their pouches and flasks, had rolled up in their blankets to " woo tired nature's sweet restorer."

On pleasant days, when shooting from my boat, I usually made a practice of reloading as fast as possible between shots, carrying an ammunition-box and loading-tools with me for that purpose; but this, for obvious reasons, I could not well do on very stormy days or when shooting away from my boat, and, consequently, I had to refill my shells at night—often when I should be sleep-ing—or else forego my morning shooting next day.

Frequently in spring continuous shooting may be

had at "travellers," as they are termed by the hunters—*i.e.*, ducks making long flights, often migrating, flying high in the air, offering few shots
under fifty yards and more over sixty. This
kind of shooting requires a good gun and heavy
loading, lots of powder, and large shot if procurable—so much powder, five to six drachms,
that it would be folly to make a practice of
using so large an amount in ordinary sport; consequently, the shells being usually loaded with
common charges, this sport must be given up, or
the shells emptied and reloaded more heavily, a
tedious job when ducks are flying thickly. With
a muzzle-loader all one needs to do is to raise
his charges a notch or two, and he is ready to
kill his farthest. Again, on the other hand,
very close shooting may often be had when available ammunition is not very plentiful. It is then,
of course, desirable to lessen our charges as much
as possible, in order to receive the greatest benefit
from our opportunities. And here, again, the
muzzle-loader has the preference.

To be brief, a man risks fewer long, wild shots
with a muzzle-loader, and consequently wastes
less ammunition, has less extra bulk and weight
to carry, as shells, loading-tools, etc.; and in boat-

shooting, if he uses *two* muzzle-loaders, can, I think, kill more game the season through than with one breech-loader, as he will frequently have opportunities to shoot both guns into the same flock of ducks before they get out of reach. I shall describe hereafter a method for loading a muzzle-loader in which the operation is greatly shortened, and much valuable time saved.

Now, because of my saying a few words in favor of the muzzle-loader, do not consider me an old fogy, or old-fashioned either. I have not finished yet. I intend to give the claims of the breech-loader an equal showing, and, I think, can still find advantages enough to overcome most of its deficiencies. And first (I will be brief as possible), they have the advantage of rapidity in loading, whereby in wild-fowl shooting, besides the advantage of always being ready for new-comers, cripples may be the more readily secured. Second, ease and quickness of exchanging loads in a gun, as in the case of the approach of geese or swan when awaiting ducks. Third, facility of cleaning. Fourth, less liability to miss fire. Fifth, safety, no getting two loads into one barrel, no need of having head or hands over the muzzle, or leaving gun loaded when not in use or when riding in a wagon or other vehicle.

Now, to sum it all up: In my opinion, for the majority of uses, the breech-loader is the superior weapon for the novice or the gentleman sportsman. For the poor market duck-hunter, if he can afford but one gun, I would advocate the muzzle-loader; he will find it much less trouble to take care of, and less work to keep loaded; he may kill a few more ducks with a breech-loader, but they will cost him enough more to make up the difference both in labor and ammunition. For boat-shooting, I would advise, where it can be afforded, the use of both guns, when either may be used as occasion demands, and the special advantages of each be secured.

We will suppose our reader to have made up his mind as to which class of guns suits him best—breech or muzzle loader. And, next as to dimensions, etc. For flight-shooting, an 8-bore is as large as is advantageous, and a 10 is sufficiently small. For a breech-loader, I should never use larger than a 10, as the cartridges for an 8 are too bulky and require too much room for transportation, and the 10, if properly loaded, will kill nearly if not quite as far. For a muzzle-loader, a slight increase in the gauge will admit of larger charges being used for flock-shooting, while the objections in the case of the breech-loader do not occur. Their weights in proportion to their gauges should be as follows:

For a 10-bore, 9 to $10\frac{1}{2}$ lbs.; 9-bore, $10\frac{1}{4}$ to 12 lbs.; 8-bore, 12 to 14 lbs. The barrels should never be under 30 inches in length, while anything over 35—and that only for an 8-bore—is superfluous, and only waste and useless metal to carry. Damascus is, I think, the best material, on account both of its beauty and superior strength. Laminated steel, if properly made, is about as good, but so many cheap and almost worthless imitations are in the market that it cannot be relied upon. Damascus cannot be cheaply counterfeited, and therefore is more to be depended on. Select barrels, if possible, having nearly a true taper from breech to muzzle. The rib should be about $\frac{1}{2}$ to $\frac{5}{8}$ inch wide, slightly concave laterally, but perfectly straight lengthwise, and sufficiently elevated at breech to prevent under-shooting. Sight-piece small and close to the muzzle.

So many different principles and methods are employed for breech-loading actions that a full consideration of their merits and defects will be

impossible here. To my own mind, the best and
most desirable is the under-lever double-grip,
on account of both its great simplicity and durability, though many of the first-class snap actions
work very freely and wear quite well; in fact,
for ordinary use are sufficiently durable; but for
rough, every-day work, nothing, in my opinion,
equals the double-grip. I have fired one over
twelve thousand shots, and though never in the
hands of a gunmaker to be repaired, it is now apparently as tight as when new. "Greener," the
celebrated English gun manufacturer, in his book on
" Modern Breech-loaders," says : " The double-grip
is considered by all practical gun-makers to be
the strongest and most durable arrangement for
sporting guns and rifles. Nothing can be more
simple or do the work better. There being a
great amount of leverage, it possesses wonderful
binding power, and when properly made and well
fitted it will last many years without becoming
loose, as it can be made self-tightening to allow
for wear and tear. It is getting more into favor
every season amongst the sportsmen at home and
in India. When guns and rifles constructed on
the double-grip plan have failed, it is attributable
to the imperfect mode of making the action. We

have seen long, heavy double rifles of 8-bore, fitted to a light breech action with bar-locks, and the metal cut away just where it was most required, being left barely strong enough for a light shotgun."

The break-off in the breech-loader should not be less than three-fourths inch in thickness, and the strap long and fastened to the stock by two screws. The locks, as quoted from Greener, should be back-actioned, thereby leaving the action stronger, and should have a fine oily feel, and give forth a sharp, clear click on cocking. (For a muzzle-loader I should prefer forward-action locks, as they are more pleasing to the eye, and do not weaken the gun any more than the old style, if as much. They should be independent, as they are less liable to get wet inside.) The mainspring should be rather stiffer than is usual in muzzle-loaders, and the hammers should have a good sweep, as some breech-loading caps require a strong blow to explode them. Rebounding locks are rather a detriment than an advantage. They are more liable to get out of repair than the common lock, cannot (from principles of construction) be made to strike a blow sufficiently heavy to always insure exploding the cap or primer; and from the sudden jerk upon the mainspring are liable to be broken in very cold weather. Another very serious objection, which I have seen occur myself, is that the cap may be driven back into the needle-holes by the force of the explosion, the hammers not holding the strikers up to them, and the working of the gun thereby for a time prevented. If the strikers are sufficiently long, however, to fill the hole completely, this objection cannot occur.

The stock should be of English or German walnut, with a strong, thick wrist, and the grain and fibre of the wood running with the angle of bend. A pistol-grip is thought by some to be an advantage. The stock should be varnished and polished. An oiled stock does not stand water well: when wet the fibre of the wood is raised, the wood is swelled, and on drying shrinks from the metal work, leaving the joints open.

The trimmings should be of case-hardened iron, with little or no engraving.

To determine the length and bend of stock required in ordering a gun, the best way is to procure a gun, if possible, which seems to "come up" to suit you, then lay a straight-edge along the top rib sufficiently long to extend to the butt-plate, and measure the distance from the underside of the straight-edge to the stock, both at the top of the rise from the wrist, or the nose, as it is called, and at the butt-plate; this will give the bend. From the right trigger to centre of heel-plate is the length of stock required. About three inches is found to be the ordinary bend, and fourteen to fourteen and one-half inches the usual length of stock required.

As to the shooting powers of the gun when properly loaded, they are dependent mainly upon the form of bore, in connection with the elasticity of the material of which the barrels are composed. It is a very foolish idea, though one, I am sorry to say, quite prevalent among sportsmen, to suppose, because some one gun is found to do very strong and close shooting, that all others made by the same maker will do equally or nearly as well. This is a very unreasonable presumption, especially in the judging of those gunmakers who manufacture to order. Their customers generally order their guns made to shoot as their use requires. One who may not be over-particular in aiming, or who desires a gun solely for wood or brush shooting, and where he seldom has to shoot far, will, perhaps, order the gun made to shoot open. Another, who may shoot nothing but ducks, may require his to shoot very close. It is obvious neither alone could be taken as a fair sample of the gunmaker's abilities.

As to factory guns in general, they are meant to be made to shoot passably before leaving the shop, and, where they fail to equal the ideal of the purchaser, the dealer usually rebores them to shoot as desired. Thus, again, as many guns are rebored after leaving their makers, their good or bad shooting qualities cannot with justice be ascribed to them. Sportsmen's ideas, too, differ in regard to shooting so frequently, that what one might call an extra shooting gun, another would consider as only ordinary. Nearly all our first-class gunmakers understand boring fully, as an essential part of their business; so in ordering a gun, if the sportsman will specify how he wants the gun to shoot, he will nearly always be suited, if his demands are at all reasonable.

Before I go further, let me explode another foolish notion entertained by many of the thoughtless ones. Because game may be killed with more certainty at short distances with small shot than with *too* large a size, or because they may sometimes happen to kill an extra long shot when using them, they have concluded that small shot will kill farther. Now, on asking their reasons for this belief, several have told me—and, indeed, one late author has published the same theory—that a small shot striking a bird, say through the lungs or stomach, makes but a small hole, which closes after the passage of the shot, thus preventing the escape of blood, and causing the bird to die quickly from internal hemorrhage; whereas a larger shot striking in the same place leaves an open hole, through which the blood runs freely, and the bird flies on frequently out of sight, or until it dies from sheer loss of blood.

My own idea is that fully nine-tenths of the game that dies solely from loss of blood or internal hemorrhage is never recovered by the sportsman; and though I admit they die more quickly when bleeding internally than if the blood flows outwardly, yet from the wound made by the larger shot, as more of the veins or minute blood-vessels are severed, more blood would escape, and the choking from internal hemorrhage would ensue full as quickly though a portion of the blood should pass through to the outside.

It is by the severe shock or paralysis of the nervous system more often than otherwise that death from gunshot wounds is produced, and this alone is the almost invariable cause of instant death in such cases. Assuming this to be the fact (and I think but few of my thinking readers will hesitate to do so, being supported in the opinion by the testimony of our most skilful surgeons), it follows that our object should be to create a shock sufficiently severe to always insure death if possible. It is well known that a comparatively slight blow in a vital point, as certain parts of the head, neck, or the immediate region of the heart, will produce a shock sufficient to cause death. A very heavy blow, or the united shock of a number of lighter blows, taking effect in less vulnerable parts, may be sufficient to accomplish our purpose. So long as we insure striking a vital place, it is obvious the larger the shot we can use the better, as, their momentum being greater, and their individual striking surfaces larger, they must consequently have greater bone-smashing and nerve-destroying effects, and produce greater shocks. Once in a while, though once too often, a stickler for small shot will assert as an argument in their favor that small shot will penetrate deeper than large ones, as their surface to be resisted is so much smaller.

But, without arguing the point, let the sceptic
put it to practical test, and if he finds in shooting at the same target of pine-wood, all other
things being equal, that he can get deeper penetration with No. 6 shot than he can with No. 1,
my faith in the certainty of things in this life
will be sadly shaken.

Having wisely chosen our shot, our desire
should certainly be to give as great a force
to those shot as is compatible with safety,
comfort of shooting, and sufficient closeness and
regularity in their dispersion; for an excess
of powder over the proper charge will
cause the shot to scatter widely. The proper amount of powder, as well as the size
and proportionate quantities of shot, we can
determine only by experiment. As to the
amount of each suitable for duck-shooting, from
my own experience and observation of the
charges used by the most successful duck-hunters of my acquaintance, I find the best proportions to be:

For a 10-gauge, 4 to $5\frac{1}{2}$ drachms powder, 1
to $1\frac{1}{4}$ ounces shot.

For a 9-gauge, $4\frac{1}{2}$ to 6 drachms powder, 1
to $1\frac{3}{8}$ ounces shot.

For an 8-gauge, 5 to 7 drachms powder, $1\frac{1}{8}$ to $1\frac{1}{2}$ ounces shot.

These charges are amply sufficient for single ducks, and will kill as far as heavier ones if the game be fairly held on. For flock-shooting or a poor marksman, more shot may be added with advantage, as giving more striking surface; and in using larger shot than No. 4, one-quarter ounce should be added to each of the foregoing charges, on account of the fewer number of pellets contained in the ounce. As good a way as any to determine the best size of shot to be used for game is to shoot at a target the actual size of the game it is intended for, and at any distance where you can be tolerably certain of striking the target, with four or five pellets, and sufficient force and penetration (which is the main point), you may be sure you will be able to secure your bird if fairly held on.

For duck-shooting with a breech-loader I would recommend the use of metallic shells. I am satisfied they shoot stronger. Paper shells are very liable to get wet and thereby spoiled, besides requiring so much room if a large amount of shooting is expected. Their extra cost, too, though an insignificant item in the minds of some, is not to be overlooked by the poorer sportsman.

I have seen an objection made to the metallic shell in the columns of the *Turf, Field, and Farm*, I believe, by a paper-shell man, to the effect that they were dangerous to use. In support of this he goes on to say that, having had one miss fire, he put the shell into a vise, and was punching a hole in the cap for the purpose of prying it off with an awl, when it exploded, and he narrowly escaped serious injury. This reminds me very much of the boy who, to discover whether his gun was loaded or not, commenced to blow in the muzzle. Seeing the hammer was down, and thinking perhaps that prevented the air from escaping, he endeavored to cock the gun with his toe, which slipped, but not until he had raised the hammer sufficiently, however, to convince him undeniably that the gun *was* loaded. Had he been permitted to live a few moments longer, *he* perhaps might have been led to remark, in the simplicity of his spirit, that "loaded guns were dangerous."

I have known two cases of shells being exploded accidentally—one a paper, the other a metallic shell—both in capping after the shells were filled. The paper shell burst, splitting the holder's thumb open, and depriving him of the use of his hand for several days. The metallic shell did no serious harm whatever, not bursting, delivering its charge rather forcibly, but luckily not towards any one. By capping the shells before filling, which should *always* be done, all positive danger from accidental explosion is avoided.

The "Sturtevant" shell I like best for the following reasons: First, the loading apparatus is reduced to a minimum. From its construction, the anvil and ejector remaining in the shell, the extra tool for punching off caps needed with all other shells is dispensed with, the rod for pushing down the wads answering the purpose. Second, they are less liable to miss fire. Third, the cost of caps, which are the same as used in the paper shells, is less than any others, excepting common muzzle-loading, and, in cases of emergency, G. D.'s may be used. Fourth, with proper care they are more durable than any other shell.

Wads (Eley's are the very best ones) two sizes larger than the bore of the shell should be used to prevent the displacement of the shot in one barrel by the discharge of the other, as frequently happens if wads of too small a size are used. Repeated firing, however, will loosen most any wad; so the sportsman if in the habit of firing one barrel more than the other should, after firing that barrel two or three times in succession, change the loaded shell remaining in the gun to that barrel, and put the fresh one in its place. Some sportsmen of my acquaintance use wads three or four (and one five) sizes larger than their guns; but this I consider decidedly going to extremes. Where one of so large a size is used it crimples, and holds even less than a smaller one. If the wads are at all thin, two should be used over the powder. One alone is apt to be blown to pieces in the barrel, causing the gun to shoot badly.

More breech-loaders get shaky in the action by being worked carelessly than from repeated firing, and when buying a gun the purchaser is seldom taught the proper method of using it, so I will attempt to describe it here. The barrels should never be allowed to drop down suddenly, bringing up with a sudden jerk, as is the favorite way with the snap advocates; neither should they be thrown back into position with a snap; that must wear the hinge excessively. But, on taking the gun down from the shoulder after firing, drop the stock inside the elbow, and hold it firmly against the body with the upper arm ; then, grasping the barrels tightly with the left hand a few inches in front of the hinge, unfasten the lever with the right, and lower the barrels down easily. Use either hand to withdraw and insert the shells, holding the gun in position with the other. The gun should then be closed in the same careful manner. The whole operation, so long on paper, can thus be performed as quickly as in any other way, if not quicker.

In loading a muzzle-loader, study to make as few motions as possible, and those short and direct. If shooting from a boat, have a large, straight rod nearly the size of the bore, with which you may push the wads down as quick as you please. A quick-loading flask, *i.e.*, one having a large feed-hole to the charger, should also be used.

It is often desirable, where ducks are flying in spurts, or cripples are to be secured, to load as fast as possible. In order to do this, the shooter should provide himself with a few thin metallic tubes (tin is good), about an inch and a half in length, with an inside diameter equal to the bore of the gun. Then,

first placing them on a level surface, he is
to push a wad into each as far as the bottom,
and, after cutting in the proper loads of shot, is
to secure each firmly with another wad. Then,
when in a hurry to load, all he needs to do,
after dropping his powder into the barrel, is to
place one of these tubes over the muzzle (guides
should be soldered to the outside of the tubes
to insure their being in the right position), and
with his rod push the contents down at once,
when capping finishes the operation. The tubes
may be refilled during the intervals between
shots.

The proper accoutrements for carrying powder
and shot are so universally known that to describe them would be simply a waste of time.
To those who may be in need of such advice,
however, I will just say, if you *will* load from the
tin canister you buy your powder in, get something else besides a *screw*-top; and, if you *must*
use a bottle to carry your shot in, try and find
one with a neck large enough to prevent the shot
from jamming and stopping it up when coming
out; and, though you may save a cap or two by
it, it is not economical in the end to carry your
caps in a box, which frequently, in fact always,

when the ducks are flying thick, you must open with your teeth if you want a cap. Carrying them in your vest-pocket is (take my word for it until you try it) full as handy.

Now, to both old and young, let me add the caution: Be careful in handling and carrying your gun. NEVER CARRY IT WITH THE HAMMERS DOWN ON THE CAPS. At half-cock is the proper position; then, if the locks are well made and in good order, it will be almost impossible for the gun to be accidentally discharged. Even at full-cock there is less danger than with the hammers down.

More of the frightful accidents with guns are attributable to this carrying with the hammers down than to any other one thing. How many times guns are accidentally discharged without serious consequences we have no means of finding out. If it could be known, I am satisfied the number from the reasons above given would exceed all others combined.

A very careless and dangerous way of carrying a gun, though a very common one with some, is to grasp it by the muzzle, with the barrels resting on the shoulder and pointing to the front. I have myself known of three fatal accidents

caused in this manner, all directly occasioned by
stumbling, when, in guarding against the fall, the
stock was suddenly thrown over to the front,
and the hammers, striking the ground, exploded
the charge.

One of the safest, easiest, and readiest methods
of carrying the gun is across the front of the
body, the barrels pointing diagonally upwards, the
fore end of the stock resting in the hollow of
the left arm, and the gun held in position by
the left hand, which grasps the wrist of the
stock. The right hand instead of the left may
be used to hold the stock, in which case the
fingers of the left simply lie behind the hammer. With the barrels over the shoulder and
the stock to the front is always a good way;
but care should be taken, especially if hunting
in company, to keep the muzzle well elevated.
In a boat, lay the gun in such a position that
the muzzle may be pointing from you, and hunt
as little with a greenhorn companion as possible.

THE ACT OF SHOOTING.

I shall suppose my readers to have at least
a fair knowledge of shooting in the field, and,
therefore, will not attempt a discourse upon the

SHOOTING.   **41**

A B C of the art. It is an acknowledged fact, however, that some of our most successful field-shots
frequently make rather poor work of shooting
wild fowl. Their usual fault is in hurrying too
much, not taking time to make the necessary
allowance for the rapidity of flight. Straight-
away shots they usually kill better than any
others, because little or no calculation is required;
but in cross and over-head shooting, where most
judgment is called for, their shot too frequently
passes behind the bird. A few brief instructions
upon the subject, therefore, may not prove in-
appropriate.

Never bring up the gun in a direction opposite
to the bird's flight, nor put it up in any way
in front of the birds, waiting for them to come
to it; but wait until they get nearly to you,
and then, bringing the gun up directly behind
them, carry it forward quickly in the exact line
of their flight, and pull the trigger without
stopping the motion of the gun. The precise
time of pulling, and the amount of space which
must be allowed in front of them and behind
the line of aim, will, of course, vary greatly in
accordance with the direction and apparent velocity
of their flight and the probable distance they

may be from the shooter. All these conditions, and the allowance to be made, you must estimate almost instantly, whilst putting up the gun, and without musing or pondering over it. This, of course, can be learned only by practice; no instructions can convey the art.

The rates of allowance vary with different people, some almost imperceptibly arresting the motion of the gun at the instant of pulling trigger, others stopping it almost entirely; therefore it will be impossible to give any precise rules by which this may be determined. To kill mallard when flying at their usual rate of speed, I myself should aim, I think, about two feet, or their length, in advance, if at a distance of thirty-five yards from them. This may help to give the tyro a proximate idea of it, though he may find in practice, for the reasons given above, his proper allowance to be either a little more or less. With experience, the hand and eye will seem to act intuitively without prompting from the mind; but it will require painstaking attention. It is not enough to toss the gun up carelessly, and to shoot anywhere in the direction of the ducks.

To make the best work of sitting shots, different rules are given by various authors, some saying to shoot at the birds as they are turned from us on the water, and others telling us to wait for the broadside chance. One thing certainly you may rest assured of, which is agreed to by all: when the birds are facing you is the poorest time.

Your position should not be too high, but about two feet above the level of the ducks, if convenient. If the flock is large and close to you, do not shoot at the nearest ones, but rather beyond, inside the edge of the flock, as many pellets which in the first case would be wasted in striking the bodies would, if delivered as directed, take effect almost entirely upon the heads and necks, most vital parts; those of the nearer ducks, if the flock is closely packed, often completely shielding the bodies of those farther off.

The ducks may be engaged in feeding, some of them tail-end up, with their heads and necks under water. It is useless to shoot at them in this position. In such a case the shooter should give a low whistle to make them raise their heads before firing. Do not start them up, as some authors advise, but shoot as soon as their heads are raised. More may be killed when on the water by following the above directions than in any way after they have started.

Do not use shot of too large a size, nor try to get too near, but give your charge a chance to spread.

In shooting over your cripples, which should be done as soon as possible, secure the liveliest one first, and try, if you can get two or three in line, to shoot them before they separate. The dead ones should be the last gathered.

CHAPTER II.

THOUGH the principles of general procedure may often be the same in like varieties of wild-fowl shooting, the different surroundings frequently necessitate the exercise of considerable ingenuity in the providing of proper ambush, or blind, as all such hiding-places are generally termed by wild-fowlers. And as it will save considerable labor to know how to set about it properly, I will devote a few lines to the subject.

The first thing to be done before building your blind is to decide upon its most favorable location; and this decision must be governed by various conditions influencing the actions of the ducks, and which you must understand, as well as the habits of your game, before you can be sure of being right. When you enter a pond, note how the ducks may be sitting, whether scattered promiscuously about it, or grouped in some particu-

lar place.   Where they are thickest they care
most to be.   On putting them out, note how they
leave the pond; they will almost invariably return
from that direction.   They seldom take a roundabout course.   Note the position of the sun and
time of day, remembering the sunny side of the
pond is best for decoys.   Note, also, the direction
and force of the wind, and its probable influence
on the ducks.   From a proper consideration of these
and various other little items, not easily enume-
rated here, 1 will suppose its location determined.
Now, if a natural blind can be found, such as an
old tree-top or roots, a bunch of bushes, or such
like, in a suitable position, it should, of course, be
taken in preference to building a new one, as the
ducks, accustomed to the object, have become
familiar with it, and, having no suspicions of
danger, do not hesitate to approach; but if such
a blind is not to be had, your next course will
be to decide upon the most suitable materials
handy for building an artificial one, and these, with
its shape, should be selected as nearly as possible
in consonance with the nature of the surroundings,
an improper selection exciting observation, and
consequently suspicion.   Take plenty of time and
build your blind well; make it look as natural

as possible, and sufficiently large and impenetrable to sight to afford proper concealment. If the ducks are liable to approach from different directions, build it to enclose you completely. A half-built blind is a nuisance.

It is certainly laughable to see a greenhorn behind a blind such as he usually builds—a few bushes stuck up to dodge around—when, as it often happens, a couple of flocks of ducks may be approaching at the same time from different directions. At first he tries to hide from both, but, giving that up as impossible, makes up his mind which of the flocks is of the most heedless disposition, or is coming most directly towards him, and so jumps to that side of his clump of bushes which affords most concealment from them. On looking over his shoulder an instant, however, his mind wavers, and, affecting his body, that, too, begins to waver, first to one side of the blind, and then to the other, as the vacillations of the mind seem to prompt it. All his motions, however, only serve to attract the attention of the ducks, and they swerve by to either side far out of reach. He now deliberates awhile, and concludes his blind is not large enough. So he starts for the nearest timber or patch of bushes to cut more brush, watching as he goes for the approach of the ducks. Just as he gets to the timber he sees a flock coming, and back he runs as fast as possible, perhaps through mud and slush, arriving at the blind as they go by, too wide of course by rods. Now he is sorry he left the blind, and remains fearful to leave, lest others may come ; but upon their coming, and again disappointing him, he fully makes up his mind (if the ducks will only stay away long enough) to get more brush ; does so, and finally succeeds in getting a half-decent blind built, about the time the ducks quit flying. If he has a dog to whistle to and bellow at, and to yank around the blind when ducks are approaching, it adds very materially to the entertainment in the eyes and ears of one who can appreciate it.

In high wild oats or flags of course no building is required. The boat, if shooting from one, should be pushed into one of the thickest bunches, at right angles with the main line of flight. Then the tops of the stalks or flags are to be struck down and in towards the boat with an oar, covering as near as possible the bow and stern, and afterwards trimmed so as not to interfere with the swinging of the gun, and the blind is complete. When two persons are hunting in company in a

RICE POND MALLARD SHOOTING—FALL.

rice-pond, it is well for one to take a stand on one of the large muskrat-houses nearly always to be found there, as by taking separate positions more shots are obtained. To build a blind in a rat-house, a large one like a small hay-stack should be selected, a hole dug in the middle with the hands and feet, and the edges then built higher with stalks of rice or flags. This makes an excellent blind, as the ducks, being accustomed to rat-houses, take no especial notice of it. It is a favorite manœuvre of greenhorns to crawl round the outside of rat-houses, endeavoring to hide, and being liable to be kicked off upon firing. I have crawled about many a one thus in my early duck-shooting days.

If the blind is to be built of small branches or bushes, they should be stuck up in the ground close together, smaller twigs entwined among them, and bunches of grass, weeds, rice, or flags scattered judiciously over and amongst them, to close all large, open spaces or thin places that the ducks might see through. If very large, bushy branches are used, they may be laid down crossing each other, with the tops turned outwards. The blind should *never be built higher than the shoulders when in an erect position.*

In cutting down a willow blind about a boat, as the common blinds are made in spring, considerable judgment is necessary. As the ponds are usually bordered with willows, it is generally easy to find a group growing in the position desired, the most favorable one being that where four trees grow as it were in the angles of a rectangular parallelogram, being apart in one direction the width of the boat, and in the other about three-fourths its length. If in such a position that the boat must be head on to the decoys, the boat should be placed between them, and the trees *felled towards the bow*, the cut ends allowed to remain on the stumps, the tops of the forward trees crossing each other on the bow, and the after-tops lying on the forward trunks. If the tops are not sufficiently leafy and dense, more branches must be cut from the neighboring trees and placed upon them, and it will sometimes be necessary to tie these branches in position to prevent their being blown off. Should the trees grow the other way, *i.e.*, the long side of the parallelogram towards the decoys, they should be felled, *those on the same side of the boat towards each other*, and branches should be added and fastened. The new cut ends of the trunks and stumps should always be covered with mud or grass to hide them from sight of the ducks. Should the forward trees be the proper distance apart, it is a good plan to wedge the boat between them, thus making it more steady and better to shoot from.

In blue-bill shooting upon the edges of our ploughed prairies and corn-fields, an excellent blind may be made by turning your boat upon its edge, and bracing it in that position by a stake or oar. Game do not appear to be at all suspicious of it. For teal and golden-eyes this plan answers nearly as well; but mallard and canvas-back are generally shy of it.

In the winter, when the ground is covered with snow, a blind made of bleached cotton-cloth, fastened to stakes stuck in the ground, affords a good concealment, and cannot be easily distinguished from the surrounding snow. A white handkerchief should be worn over the cap or hat.

Great quantities of ducks are often killed in the air-holes about freezing-up time. Long after the feeding-ponds are entirely covered with ice the ducks remain feeding in the corn-fields miles from the river, to which they return to roost at night, in holes which they keep open during the severest weather by the warmth of their bodies, and by keeping the water constantly in motion. It is not unusual at such times to kill over a hundred during the day. One of the best blinds for this kind of sport is made of ice. It should bè cut in cakes, the size of which should be proportioned to its thickness, and these should be placed on edge or end in the proper form. If the ice is thin, say three or four inches thick, and the day cold, shallow grooves should be cut in the bed-ice, and the ends of the cakes placed therein. Water should then be poured about them, and the fine ice made in chopping packed in beside them, which will quickly cement together, holding the cakes firm and upright. Old ice or ice mixed with snow is the best, as new ice, if thin, is generally too transparent; but, if white cotton cloth be hung inside the new ice, it makes the blind all that could be desired.

Another and perhaps the very best blind that can be made for air-hole shooting is the sunken box, not the battery described under the head of canvas-back shooting, but a deep box of pine, almost forty inches square on top and fifteen inches on the bottom. On account of the difficulty of sinking it, it should be as small as convenient, and the sides made tapering from top to bottom; or like the figure given below, which I think is the better plan; the lower part being as deep as from the knee to the sole of the foot, the upper part sufficiently deep to completely hide the body of the shooter when in a crouching position. To sink the box, a square hole, a trifle larger than the outside of the larger box, is cut

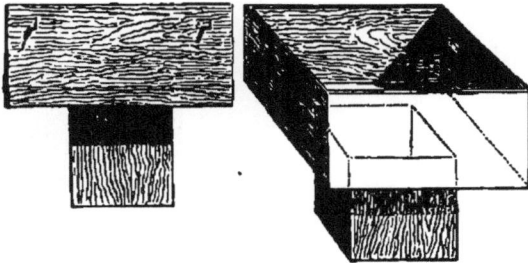

in the ice where the box is intended to be placed, and the box then sunk to the desired depth by loading it sufficiently with water. It is now fastened in position to two stout poles, about twelve feet long, which have been previously pushed under the surrounding ice, one along either side, and touching the box. The water used in sinking is now bailed out again, and, after hiding the edges of the box with pieces of ice, it is ready for occupancy. When the ice is not sufficiently strong to hold down the empty box, this plan must be given up, and the box kept to the desired depth by stones or other heavy weights.

CHAPTER III.

ONE of the most important requisites to insure success in wild-fowl shooting, and more especially in the pursuit of the deep-water varieties, is a suitable flock of decoys. They may be made in a multitude of ways, and of several different materials, each of which has its peculiar advantages, but at the same time its corresponding defects. The principal objects to be attained by all, however, are naturalness, or a sufficient resemblance to the species they are intended to represent, with the proper shape necessary to enable them to ride in an erect position during the heavy blows they are often exposed to. This last desideratum is often partially and, I might say, entirely overlooked in the desire to make the decoys as light as possible, and of such shape as to take least room in transportation. With such objects in view, would-be inventors have tried a variety of methods in making them,

55

and though certainly accomplishing their object in this respect, have failed most decidedly in the main thing needed. One of them gave us rubber decoys for the modest price of thirty dollars per dozen. They were hollow, with a tube attached, through which, when needed for use, they were to be inflated with the breath, which being ejected by compression when ready for transportation, they could be packed in very little space. They would float remarkably light and airy, a property, though contrary to general supposition, not at all desirable, as causing them to roll sidewise in the least ripple, a motion the natural ducks never make, even in the roughest weather. A shot-hole ruined them, and as the rubber soon began to crack after but little usage in a hot sun, they soon proved a failure. Decoys of metal, too, were tried, both of copper and tin, made to be taken apart, and the several parts nested together for packing; these, besides being very expensive, were proved to be comparatively no better than the rubber ones, for reasons very obvious to the knowing ones, but which the "greenies," who want everything new, could not see until they had paid their money to find out.

Decoys made of wood (not the things one usually finds for sale in the gunshops, where they should be allowed to remain, but as constructed to use, according to reason and with a proper appreciation of the thing needed) are preferable to any others. Having had some little experience in their manufacture as well as their use, and having the satisfaction of seeing my own used as models by better hunters, I will describe them as I think they should be made; willing at the same time to yield due deference to the opinions of others.

My principal object has been to secure the best shape possible to prevent rolling, and to ensure with least extra weight an upright position at all times when in use. How I have endeavored to do this will be better understood from the annexed cuts, representing outlines of the

Decoys.

decoy, than by any explanation I could convey in words.

White cedar and soft pine are undoubtedly the best woods for decoys, on account both of their extreme lightness and ease of cutting. Pine perhaps is better for heads, being less easily broken, while cedar is the most durable. The timber should be well seasoned and free from knots and sap. For ducks, 2×6 inches is the proper size, but for geese larger timber is needed.

The timber, being planed on one side and sawed in proper lengths, is next cut around on its edge, according to a pattern representing a horizontal section of the decoy intended. Two pieces are needed for each decoy, which must be gouged out to the proper thickness, thus making the decoy hollow. The head (which has been previously shaped) is fitted and fastened to the top part by a screw from beneath, and the two parts, being roughly hewn into shape in conformation with a side pattern, are, after being nicely fitted, glued or otherwise cemented firmly together, and the decoy rounded and finished smooth. After being thoroughly sand-papered, it should be wet slightly all over so as to raise the grain of the wood, and when dry should be again sand-papered. If the decoy be washed over with a thin dressing of shellac, it will prove much more impervious to water. This should be done before painting, and no varnish should be put on afterwards, as it makes the decoy too glaring when in the sun. When thoroughly smooth, a heavy coat of priming should be put on, of some neutral tint that will not show too plainly through the coloring coat; all of which should be mixed with raw oil, and without an artificial drier. The priming should be allowed to harden thoroughly before the colors are put on. No priming is used on many of the decoys for sale in the gun-shops; consequently, they soon become water-soaked and heavy, and the colors indistinct. Artists' tube colors should be used, being more lively and durable than common paint, and costing but little more; and the nearer the painting resembles the coloring of the natural duck the better. A small brass wire staple or piece of leather is to be fastened to the lower part of the breast, to attach the line to. A piece of lead, about four ounces in weight, formed as shown in the figure, should next be screwed on to the bottom lengthwise, like a keel, and the decoy is complete.

For shoal-water duck-shooting, flat-bottomed, hollow decoys, of two and one-half inches in thickness, answer fully as well, as the water is seldom rough.

Each decoy should be provided with a separate line and anchor, which last should be of lead, if convenient, as it is less liable to scratch the paint from decoys than anything else. This need never exceed four ounces weight. The line should be what is known as "sixteen thread" seine twine, about one-tenth inch thick, of a length adapted to the depth of water, and attached to the staple or leather in the breast of the decoy. Instead of winding the line round the neck of the decoy, as is often done, the proper way is to wind it tightly round the middle, which may be done in much less time, an item of importance when taking up decoys in a heavy wind. And in setting them out again, instead of unwinding them turn by turn, the decoy should be taken by the head in one hand, and the lead thrown with the other to the place desired, the turns coming off towards the tail as the lead is thrown. A large flock of decoys may be set out in this way in a remarkably short time.

In this connection it will, 1 think, be well to give a few directions as to the management of the boat when taking up decoys in a heavy "blow." If you remain in the stern, you will find it very hard to keep your boat head to wind; when stooping to pick up your decoys, it will whirl round, and you will have some work to turn it back again. Therefore, stand in the bow, with your knees braced against the bulkhead or sides of the boat, and paddle bow first as usual. By so doing the boat will never of itself turn the wrong way, and you may pick up your decoys in a short time and with comparatively little labor, when it would be impossible in the manner first mentioned.

Always pick up your leeward-most decoys first, and, just before stooping to grasp each one, give the boat an extra stroke ahead to keep up its headway whilst winding the line. If you erroneously commence at the windward side of the flock, many of the lines will invariably become entangled in winding up, when those of the windward decoys must often be pulled over those nearest to leeward; and in the event of the boat's drifting back upon them and bunching them together, as will unavoidably occur if the decoys are placed as closely together as they should be, before the snarled lines are separated and wound up the boat may have been blown to leeward many yards, occasioning hard pulling to bring back again, besides the confusion it has made.

A favorite way of making decoys with some of the old sea-coast "gunners" of my acquaintance (though I never thought much of it and have never seen such used in the Western country) was simply to cut them ' *in outline* of inch •boards. These were fastened one at either end to short boards, termed floaters, about two feet long and six inches in width, by pins inserted in holes bored in the under edges of the decoys, which, being loose, left them free to turn sidewise with the action of the wind and waves. The anchor-line was fastened to the centre of the floater; and when not in use the decoys could be lifted from the pins and be packed in comparatively little space. They seemed to work first-rate, especially in coot-shooting, though I should much prefer the full-sized hollow decoy, notwithstanding the additional packing-room required. One *special* advantage they undoubtedly possessed—that of being easily and quickly made.

Another, and perhaps the best decoy for coot-shooting, is made as follows: A piece of pine-board, or cork is better if procurable, is shaped for the bottom of the decoy, and to this is nailed, at right-angles and lengthwise, another piece of board, cut to represent a vertical section of that portion of the decoy above water, the under edge being left straight to fit the bottom piece; pieces of flat barrel-hoops, or similar elastic material, are now bent over. the top crosswise, and fastened to the bottom board about three to four inches apart by tacking, thus forming a framework for the decoy. This is now to be covered smoothly with strong cotton cloth, and the edges pinned securely; the head, which is made of wood, being fastened in proper position on the edge of the vertical board, and the decoy, after being painted and thus rendered perfectly waterproof, is complete. Ballast, however, is usually added to keep them erect in rough weather. These decoys are generally made three or four times the size of the natural duck for greater show, and are a great advantage over the life-sized wooden ones on this account, coot being uncommonly foolish ducks, so much so that "silly as a coot" has become a frequent expression of the coast-gunners when speaking of a light-headed or tipsy person. This pattern of decoys 1 would not recommend for Western gunners, unless it be for goose shooting, as they are too large and clumsy for convenience in carrying.

A few years ago a man named Woodsum (if I remember the name correctly) got out a patent for a flapping decoy. A board, which served as a floater, had a hole cut through it the size of the decoy, and in this the decoy (which was made like any common wooden one) was placed and fastened to the board by pins running into its sides, and serving as hinges upon which the decoy tilted easily. Wings, formed of wire and covered with cloth or other similar substance, were hinged in position, and the decoy anchored in the usual manner. A line leading to the blind was so fastened to the decoy that upon its being pulled the forward end was raised upon the hinges to a nearly erect position, similar to that of the live duck when flapping its wings, and the wings were elevated at right angles with the body. It was quite an ingenious contrivance, and helped considerably to attract attention to the decoys, especially on dark, calm days. On such days, if a string be tied to a common decoy, by pulling it a ripple is occasioned or a motion made among the decoys, which will prove of considerable advantage. A short whistle or slight

noise of any kind sufficiently loud for the ducks to hear without alarming them may direct their attention toward the decoys, and so prove the means of turning them; but for such ducks as may be called readily, of course an imitation of their call-note is better.

Thirty decoys, at least, are needed for canvas-back shooting, and as many as two hundred and fifty are often used—the more the better. For mallard shooting, twenty wooden ones are sufficient to carry, as dead ones may be stuck up at any time. Though a few canvas-back may properly be used with mallard decoys in mallard or redhead shooting to increase the show of the flock, it is seldom, or I may almost say never, desirable or advantageous to use the mallard decoys for canvas-back, as they do not feed together and have no desire to associate. For Western duck-shooting, but three, or at most four, different varieties of duck decoys are needed, all other ducks decoying to some one of these kinds as well as to their own. Mallard and canvas-back are the kinds most especially required, while redhead and blue-bill decoys may be advantageously used. And here I will endeavor to arrange the decoys as needed for the different varieties.

For Mallard—Mallard decoys; a few redhead and canvas-back beneficial if the flock is small, and especially on overflowed prairies.

For Canvas-back—Canvas-back decoys; redhead and blue-bills added to any extent advantageous.

For Redhead—Redhead decoys; with mallard in shoal water, and canvas-back and blue-bills in deep water.

For Blue-bills—Redhead and canvas-back decoys with blue-bills (blue-bills alone do not show well).

For Sprigtail Teal and Gray Ducks—Mallard, or mallard and redhead decoys.

For Widgeon—Any or all kinds; mallard best, especially in shoal water.

Spoon-bills—Mallard decoys.

Wood-duck—Mallard decoys, though these do not decoy well to anything.

Other small deep-water ducks, to deep-water decoys. When ducks desire to come into any particular place, any decoy may help to quiet their suspicions of danger, and would then be of advantage, though of little use in other places.

To illustrate the crude ideas of some individuals in regard to decoys, I will relate a little incident. I was paddling up a certain river in company with my hunting partners one day during spring duck-shooting, when, a heavy rain-shower coming up, we took refuge under an old warehouse in the small town we happened to be passing, and while there were visited by several of the town inhabitants, to whom strangers were an especial attraction. Amongst the rest was a sporting New Yorker, dressed up in fancy shooting costume, and followed by a retinue of " saloon bummers and dead-beats," by whom his every wish was anticipated and his money most eagerly sought for. I found on conversing with him that he had come to hunt ducks, and certainly he had chosen a good place to find them, though skilful hunters were not to be had then to accompany him; in fact, when going down the river during the fall before, our cook sold to one of the town fellows the first decoys ever owned there.

We had some idea of stopping there to shoot awhile, and so I enquired of him what the prospects were, if ducks were plenty, etc. " Oh! yes," said he, " there are plenty of canvas-backs, but they are fearfully wild, and won't decoy worth a d—n." " So, then, you use decoys?" said I. " Yes, indeed; brought them from New York with me; you can't do anything without decoys, you know." I of course agreed with him there, and asked him what kind he used. "Well, I've got mallards, canvas-back, redheads, sprigtails, blue-bills, and teal," he answered rather consequentially. "Why, you have got a variety certainly," said I, somewhat surprised. "How many have you altogether?" "*Thirteen*," was the reply. I didn't ask him what the odd one was, being entirely satisfied as to why the canvas-backs were so "fearfully wild," and we afterwards found they decoyed to a nice little flock of about seventy of the proper kinds entirely to our satisfaction.

In the chapter on "Midday Mallard-shooting," I shall give a full description of the "setting up" of dead ducks in shallow water, and so will omit it here. In canvas-back and other deep-water duck-shooting, however, as it is often desirable to increase the show of the decoys, dead ducks may be fastened to them by a short line, allowing them to float some five or ten feet behind the decoy. The line should be fastened to the neck of the dead duck, which should be placed on its breast on the water. The fact of the heads not being in sight makes no material difference, this absence being probably considered by the live ones as due to the position of feeding. A small flock of decoys may be patched up in this way to make quite a creditable appearance. In cold weather, when there was no danger of the ducks being spoiled by keeping a few days, 1 used to often leave fifty or sixty, when I had them at night, covered up with leaves and brush, near my shooting-place, to use for decoys next day, and these, with the wooden decoys I usually carried, were generally sufficient to allay all fears entertained by the suspicious ones.

Live tame ducks make probably the best decoys to· be had for mallard and black-duck shooting, but they are such a nuisance to take care of and transport that they are seldom used in the West. It would almost seem as though they took an especial delight in seeing their kindred killed, from the continuous calling and quacking they keep up whenever a flock of wild ones come in sight; and they seldom call in vain, for on the wild ones hearing them they immediately turn and come in. The young wild-fowler, when shooting over live decoys, should learn to imitate their notes as nearly as possible, an accomplishment which will prove of decided benefit to him when shooting without decoys or over wooden ones.

It is often a great advantage, when shooting over wooden decoys, to have a live duck to throw in the air when wild ones are approaching. She should be secured by a light, strong line, of from fifteen to twenty-five yards in length, to the blind to prevent escape, and should be blind-folded by a hood drawn over the eyes; then, not being able to see how far she has to fall after being thrown up, she will spread her wings and allow herself to drop gradually with feet extended. as is the usual manner of ducks when alighting. The attention of the other ducks being attracted toward the decoys by her motions, they come in without hesitation. In the absence of a live duck, dead ones may be thrown up to attract attention, but do not answer quite as well, as they fall too quickly; for this reason should not be thrown too high, but rather in a nearly level direction.

When shooting teal or mallard in very shallow water with but few decoys, lumps of mud, pieces of bark, or bunches of brush of the proper size may be judiciously employed to deceive the ducks. They should be mixed with those decoys nearest the blind, but never outside the wooden ones. I have known ducks to decoy all day to a little rough patch of ground left bare by the melting of the ice along the main shore. Of course they would discover their mistake before alighting, but would .dart near enough to afford quite fair shooting from a blind near by.

Rather a cruel method, perhaps, but one attended with great success in wild-goose shooting, is, on securing a wing-broken one, to fasten it to a stake a short distance from the blind, when it will call most vociferously on seeing others approaching or passing by, who are almost certain to come if within hearing distance. Geese should be set up for decoys as fast as killed. If shooting at an air-hole in the ice, stick their heads under their wings, and set them up near the edge of the hole.

An excellent decoy for swan-shooting (they decoy very readily) is an old white shirt drawn over a bunch of brush, the sleeve being supported by a branch or stick in the proper position, forming the neck and head. A single one, if thus set out in their feeding or roosting ponds, will answer nearly as well as a dozen, but for travelling birds more are needed.

As to the position and shape necessary to arrange the decoys in respect to wind, I shall describe that in reference to battery-shooting, under that head. For point-shooting, shooting from a blind on shore, or in the edge of the willows from a boat, a few hints may be welcome.

With the *wind off shore* a very good way for .shoal-water ducks is to set them out lengthwise with the shore, rather thinly scattered, immediately opposite the blind, and grouped together, as it were, in two separate flocks at either hand. The open space opposite the blind should not be more than ten to fifteen yards wide, with perhaps five or six decoys in it, and the main flocks about thirty yards from the blind, no decoy being more than fifty yards distant. By arranging them in this manner, the ducks are allowed to come in between the two flocks, and drop into the open space instead of alighting outside the flock, as they often do when the decoys are improperly arranged. For deep-water ducks, three or four decoys as tolers may be set out to leeward, sometimes one hundred yards or more from the blind; but if so placed for the shoal-water varieties, they will frequently alight with them instead of coming on to the main flock.

With a *side wind*, the habit of the deep-water ducks is to alight with the middle or windward decoys, while the shoal-water varieties seldom pass over them, but usually alight with the more leeward ones; place your decoys accordingly.

But the success in decoy-shooting often depends more on their position in reference to the sun, if it be shining unobscured, than as regards the wind. This fact, I have often observed, is entirely overlooked by the great majority of duck-hunters. The position of the sun is seldom for a moment thought of, and, if at all, only to avoid its shining in the face when shooting, in the location of the decoys. There are so many things to be observed and considered in the selection of the position for blind and decoys, that no absolute rule can be adopted to fit all cases. Circumstances will not always allow of it; but, as a rule to be observed when conditions permit, remember to so place the decoys that *the sun may shine on that side of them from which the ducks approach.* They will thus attract attention, and be much more readily seen than if the shady side is presented. This is a secret of success in duck-shooting understood by very few amateurs, but well worth knowing. A thorough knowledge of these little things marks the difference between the lucky man and the unlucky one.

Many are the different rules given by the would-be thought knowing ones as to the best time for shooting at ducks over decoys, the most common one being to "wait until they are just in the act of alighting, and then give it to 'em." Others, who understand plover-shooting better than wild-fowling, say, "Wait for them to double." These rules may do very well in their practice, but in mine I have always found the best time to shoot was not to be decided by rule. The numbers of the ducks, their manner of approach, their species, and various actions, whether suspicious or otherwise, should influence the decision as to the proper time. And as these conditions are constantly changing, no one rule will apply.

If single ducks or pairs come in, where is the need of waiting until they are ready to alight? They may see something to alarm them, and, instead of alighting, sheer off. Besides often losing them in this way, much time is lost in waiting; and perhaps others that might be coming arrive just in time to be frightened by the wild shots made at the retreating ones, and thus two chances are gone. No; just as soon as you are satisfied they are within easy killing distance, kill them if possible. How much better it looks to see a man kill his pair prettily when flying over or by his decoys, than to wait until all headway is stopped, and then shoot as though at a sitting mark! If a small flock comes, watch to get two or three crossing, and as soon as you do, shoot; be ready at the same time to use the second barrel. When a large flock comes in, if you are satisfied they will alight, let them do so, and wait until you get several in range, if possible, before firing; but never give a single duck the chance to get away, after his once coming within thirty-five yards, without doing the best you can to prevent him.

How well I remember my old partner, Joe Carroll, the best duck-hunter by all odds I ever met! What a slim chance a duck had for its life after once approaching him within gunshot! We were shooting together from an ice blind at the edge of an air-hole one day (one of our big days too), and the ducks were coming almost continually. We had decoys and dead ones stuck in front of us, and almost every flock that came along darted to them. The blind, however, was made of new ice, and, being to a considerable degree transparent, they could see us, though rather indistinctly, through it—plainly enough, however, to make them a trifle suspicious, and to want to circle round and look at it awhile before making up their minds to alight. Though they were coming all the time, I was at first inclined to wait until they came near enough in front of us and over the decoys, the more especially as the blind was small and not easy to turn about in. I soon gave up this idea as a flock came in, and circling behind us swung in within easy gunshot, and Joe, who was watching them as they circled, jumped up, and, turning half-way round, killed his pair prettily. "Why didn't you wait until they came round in front, so that we might both have a chance?" I asked. "How did you know," was his Yankee reply, "that they were coming round in front? And even if they were," said he, "a duck killed behind us is as good as one in front; and when they are coming as fast as they are here, it won't do to lose time in waiting. I supposed you were ready and watching them when I jumped. I gave you the word when I got up. Never wait," said he, "when they once come within easy killing distance in a situation like this." And this advice I have made it a point to follow, and, I think, much to my own advantage.

Mud-hens often cause the duck-shooter consider-able annoyance, especially in blue-bill shooting, as the ducks, instead of coming to the decoys, often dart down to and alight with them. They should always in such cases be driven out of the pond if possible.

A large bag (a coffee-sack answers admirably) is the best thing to carry duck decoys in on land.

CHAPTER IV.

.

THE size and shape of a paddle-boat proper for duck-shooting on inland waters must depend, of course, in a great measure, upon the locality intended for its use, as a small, low-sided boat, such as would be sufficiently large and safe for narrow rivers and ponds, might be entirely out of place and even dangerous on larger waters; while one adapted to the latter would obviously occasion more trouble in the finding of a sufficient concealment than was necessary on the smaller streams if a correspondingly fit boat were used. It follows, then, in regard to the size, that it should be as small as possible compatible with safety and the capability of transporting any needed amount of freight. As to its shape, it should be so formed as to insure greatest speed with sufficient steadiness and seaworthy qualities, and of such material and substance as to combine least weight with satisfactory durability.

I shall first describe a boat such as built for comparatively large streams (the Illinois River, for instance), capable, when managed by an experienced person, of withstanding any weather to which the Western hunter is likely to be exposed—such as I myself have made use of more than any others, and which, I am satisfied, is the best shape for general use on Western waters, and not too large for novices anywhere.

Now, the building of a paddle-boat is not so simple an undertaking as many of my readers may suppose; in fact, it is almost an art, and simply giving the dimensions of the finished boat proves of little value, for, not knowing how to set about the building of it, the novice is as far from its possession as though he had never heard of it. I shall therefore endeavor to so explain the *modus operandi* as to enable any one having a sufficient knowledge of the use of tools to build the boat as it should be; and, to make my instructions more clear, shall refer my readers to the drawing on the following page.

The materials—clear white pine for sides, seats, bulkhead, and bottom, and straight-grained white oak for stem, stern-post, ribs,

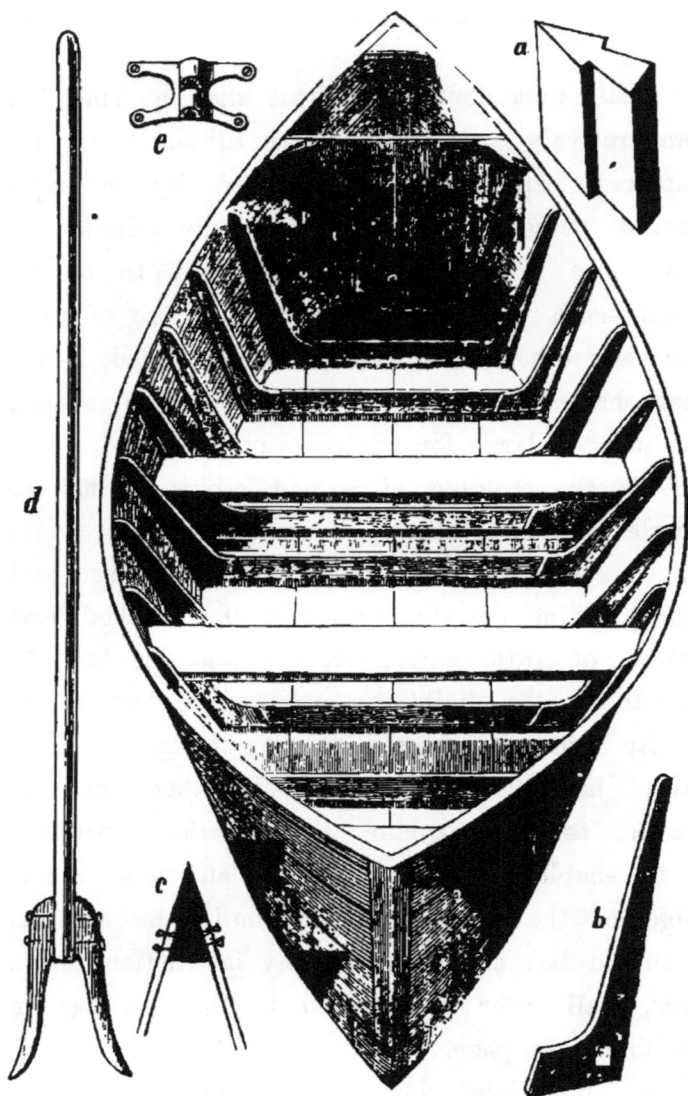

INSIDE VIEW OF BOAT.
a—Stem. *b*—Rib. *c*—Stem, showing attachment. *d*—Mud-stick
or pusher. *e*—Oar-lock seat.

knees, etc.—being collected, the first thing usually done is to "get into shape" the stem and
stern-post, the angles of which may be ascer-
tained by drawing an isosceles triangle, which shall
have for its base the width of the bulkhead at top
(18½ inches), and for its sides the distance from
front side of bulkhead to extreme point of bow (27
inches); the angle at apex of the triangle will then
be that proper for stem. The stern-post will
need to be a trifle more acute. Each should
measure 4½ inches through from front to back,
and be cut as per figure, 2½ inches back from
apex, and ½ inch deep, to receive the side-boards.

The side-boards, which should be 16 feet long,
16½ inches wide, and from ⅜ to ½ inch thick,
are next shaped in the following manner: At
11 inches back from bow, and 10 inches forward from stern-end, measured on the under
edge, cut off the end, up to the near upper
corners of board. This gives the slant of stem
and stern-post. Now, if the boards were bent
around in shape and with proper flare, the rake
of the bottom would be found entirely too
great; so to remedy this we cut away from the
under edge of the side-board a shallow arc,
which, commencing at the lower corners, rises

in a smooth curve to $4\frac{1}{2}$ inches at centre. The top edges must also be cut down at either end to a depth of 2 inches, diminishing in a gradual curve to middle of board, otherwise the bow and stern would "cock-up" too much, increasing the difficulty of paddling in a strong side wind.

The bow-ends may now be screwed to the stem (see c in the figure), care first being taken that they are evenly placed. The bulkhead-board (the bottom edge being even with the inside lower edge of sides) is next nailed or screwed in. This should be $18\frac{1}{2}$ inches in width on top and 11 inches wide on bottom (back side), and placed perpendicular with top edge of side-board, 27 inches back (measured along the side) from point of bow. The stern-ends and stern-post are now screwed together, and temporary braces put in to secure the sides in proper position to receive the ribs. The width of middle set should be, top, 39 inches; bottom, 26 inches—inside measurements. Light cords are now bound round outside of the boards to draw them in place where they are inclined to spring off too wide, and the stern-seat brace (permanent) put in. This, made of pine and similar to the bulkhead, should be 12 inches wide at bottom, at top proportioned to height, and placed square, with the top of the side-board, 30 inches forward from long point of board, measuring along its edge.

The curves are now trued by two other sets of braces, which must be regulated by the builder, the bow end being left rather more full than the stern. All is now ready for the ribs and cross-bars, nine sets of which should be used, placed at equal distance, about thirteen and a half inches apart. The ribs, as before mentioned, should be made of oak, half an inch in thickness, and two and a half in width, where they join the cross-bars, and from here are tapered (see *b* in the figure). The cross-bars should be of oak, half an inch thick, and one and a half inches deep or wide. Each rib must be accurately bevelled to fit squarely in its place; the cross-bars being cut to their proper lengths, and the flare of each set of ribs being determined by fitting, they are screwed together,* and fastened in position. Care must be taken that the bottom of each cross-bar shall be placed even with inside edge of the side-boards.

The whole is now turned over, bottom upwards, and the outside edges of side-boards planed

* Screws No. 9, 1¼ inch.

straight and smoothly in line with cross-bars, and the edges afterwards painted to receive the bottom. A straight line should now be drawn from stem to stern, and, if it is found on measuring that one side-board has sprung more than the other (not often the case, however), it must be brought back to position and secured when screwing on the bottom. The bottom boards, five-eighths of an inch thick (not more than three should be used, and two is better, if they can be obtained of sufficient width), being screwed firmly in place to the cross-bars, the bulkheads are next roughly trimmed around on their outside edges in line with the sides, and fastened to them firmly by " fourpenny " nails driven along the edge, about two and a half inches apart, the heads of which should be slightly sunken in the wood.

After finishing the bottom smoothly (not rounding the edges), the boat may be turned over, and the seats put in; one near the stern, with its forward edge resting on the permanent brace before mentioned, nine inches wide, and parallel with bottom of boat; a second at eight inches forward of the middle, and the third midway between the two others. These two last should be seven inches wide, one inch thick, and placed six inches above the bottom of the boat. The middle seat may be fastened by hooks, so that it can be taken out and replaced again at any time, if desired. This seat is particularly useful when two persons go together, when both may row, if two sets of oarlock-seats are provided, and these should be placed, for an ordinary person, eleven inches back from the nearest edges of seats.

Hand-holds are next in order ; these small angular pieces should be of one-inch oak, and fastened one at either end in the angles formed by sides and stem and stern-post. Two small angle-pieces should also be screwed in at the back corners of bulkhead and sides, to add to finish, and also to grasp when drawing the boat upon shore. After receiving the waling—a semi-oval strip of hard wood half-inch thick and one inch wide—around the outside upper edges, as a protection against wear to the sides, the boat is ready for its first coat of paint, which should be of white-lead, mixed with raw linseed-oil, and colored a light brownish drab by the addition of burnt-umber and lampblack. No artificial drier should be used, as it causes the paint to scale and crack when exposed to the action of water.

When the priming has been allowed to harden thoroughly, strips of sheet-zinc should be bent on and tacked smoothly around the edges of the bottom, from the bow back to midships. This will . protect the boat from damage whilst cutting through thin ice, and will save wear in various ways. It should ex tend up the sides about four inches at the bow, but farther back it may de- crease gradually to half that width. On the bottom it should lap about two inches. The putting on of zinc is so simple an operation that an ex planation is unnecessary. Another fin ishing coat of paint, and the boat is complete.

Side View of Boat.

The oars, which are usually cheaper bought than made, should be seven feet in length, and bound round with leather for about eight inches where they rest in the oar-locks. A light paddle about nine feet in length is also necessary, for paddling when standing up, or for pushing in shoal water or through brush. The oars should be free in the oar-locks, not pinned or hinged to them. Nothing about a boat looks more "green" or old-fashioned to me than to see an oar pivoted to a long iron pin which sticks up from the side of the boat, and is continually catching upon weeds and brush, and yet I know several first-class duck-hunters, in other respects, who use them.

To run easiest, excepting when very heavily loaded, the boat should be so trimmed * that the bottom at the bow shall be slightly out of water; then, instead of ploughing through, it will glide smoothly over the surface.

A very good style of paddle boat for small streams and ponds may be cheaply and very quickly made as follows: The bottom of inch pine is first got out in shape, thirteen feet long and twenty-five inches wide at centre, tapering in an easy curve to each end, both ends alike. To the bottom are securely fastened, at right angles, equidistant from each end, and six feet apart, two braces of pine, one and a half inches thick and nine inches in height; at the under side, equal in length to width of bottom where they

* Loaded in such a way.

join, and at the top flared five inches on each end.
The stem and stern-post are now nailed in
position, at angles of about sixty degrees, and
the proper rake (about an inch and a half)
being given to the bottom (by moving its sup-
ports either way), its edges are bevelled all
round to fit the sides; the side-boards, one inch
thick, also of pine, are bent round and nailed
in place. The lower edges are now planed
smooth with bottom; the upper ones cut to the
proper height; the braces hollowed considerably
on their upper edges to save weight, and after
painting the boat is ready for use as soon as
dry. No provision is made for rowing, but the
boat is propelled by pushing or paddling from a
kneeling or sitting position.

This boat is particularly suitable for teal, wood-
duck, and mallard shooting in the fall, but is not
large enough to carry many decoys, and it takes
in water so easily when fastened in a blind broad
side to a good breeze, that it is unfit to use for
canvas-backs, or, in fact, for spring shooting of any
kind in open water. The novice may find it
rather unsteady to shoot from at first, but that
is due rather to his mismanagement than any fault
of the boat. One accustomed to small boats need

have no fear of using it on reasonably smooth waters.

It sometimes happens that boards of sufficient width to build a boat, as at first described, cannot be easily procured. In such a case, strips of weather-boarding, or "siding," as it is called out West, may be made to take their place. The operation of building is then quite different. The first thing to be done is to prepare the bottom, and fasten the principal ribs, bulkheads, stem, and stern-post in position; then the siding is put on, commencing at the bottom in lap-streaks, copper or clout nails being used to fasten the laps, and the remaining ribs and seats afterwards added.

The sculling-float mentioned in the chapter on wild-goose shooting is rarely used in the pursuit of ducks where they are to be found in any considerable numbers. So much time is lost in the necessarily slow approaches, that more ducks are usually to be killed with less labor and more sport in some other way. But where wild fowl are scarce or appear only occasionally, the case is different. Time is then of less importance, as one or two favorable shots are, perhaps, all that can be reasonably expected during the day, and these can be obtained with most certainty by use of the sculling-float, provided, of course, the sportsman thoroughly understands the methods of operating it. This, however, requires a much greater experience than many of my readers may suppose, and even the operation of sculling whilst lying upon the back will be found exceedingly laborious and awkward to the beginner. The operation of building the sculling-float is so similar to that already described for other boats, that I will not weary my readers with a repetition of details. Its dimensions are as follows:

Length over all, eleven feet six inches.

Length on bottom, ten feet seven inches.

Width, five feet from stern, on top, three feet two inches.

Width, five feet from stern, on bottom, two feet three inches.

Width, at stern, top, two feet three inches.

Width, at stern, bottom, one foot nine inches.

Depth, one foot one inch, or thirteen inches.

Slant of stern, three inches.

At eighteen inches from stern-end the bottom rises quickly toward the stern to the height of one and a half inches, and a scag of inch oak is put on along the centre of this slant, running on its under edge in line to meet the bottom where

the quick rise commences. The object of this rise
is to prevent dragging water, and making too
rough a wake, which might alarm the game. The
scag helps to keep the boat's course steady and
direct. The bow is covered, to a distance of four
feet six inches back from the stem, with a
wash-board of quarter-inch pine, which also extends
six inches in width around the sides to the stern
to prevent shipping water in rough weather. The
sculling-hole, which should be lined with leather
to deaden any noise which might otherwise be
produced by sculling, is placed six inches to lar-
board of centre of stern, and seven inches above
the bottom. A plug of wood fitting water-tight
is to be kept in the hole when not being used.
Provision should also be made for rowing when
desired.

It is always advisable, when it can be done
conveniently, to pull your boat up and turn it
over on the bank when you come in to camp
at night, otherwise in the morning you may find
it coated on the inside with a thick white frost;
or, if it has stormed during the night, partly
full of rain or snow. If not pulled up, it should
at least be fastened in some way to prevent its

floating off in case the river should rise during the night.

A large sponge, such as is commonly used for washing carriages, should always be kept in the boat, to dry it if necessary, or to wash away stains of blood.

CHAPTER V.

A WATER-RETRIEVER is described by many of our best sporting authors as a cross-bred dog. All agree that strong powers of endurance, an unwearied, persevering disposition, speed in swimming, nose, and an acute readiness of apprehension are the essential qualities to be sought for. But as regards the particular breed or cross combining most perfectly these manifold requirements, their ideas differ widely. For my own part, 1 have had no experience in the breeding of dogs, and consequently do not consider myself capable of expressing a decided opinion. 1 have seen superior retrievers of various breeds, and to me it seemed their exceptional excellence depended more upon their individual love of the sport, and their great experience, than upon any peculiar characteristic of their breed; and their willingness to undergo hardship and fatigue was

to be attributed not so much to their actual
powers of endurance as to their ardent love for
retrieving, the charms of which outweighed all
their antipathies.

I am satisfied, from my own experience, that a
hardy, well-bred setter is as useful as any other
need be for retrieving in Western waters. By well-
bred I do not mean the silky, thin-haired, nar-
row-chested, and slab-sided animal so fashionable
nowadays; but a long, thick coated, deep-chested,
round-ribbed, and broad loined dog, capable as
well as willing to do hard work. Such a one, if
rightly taught and properly managed, will never
refuse to go where any dog should be sent; and
very few cripples will ever escape his untiring
activity and perseverance. The fact of his excelling in the pursuit of other game cannot detract
from his usefulness as a retriever, but, on the
contrary, is a positive advantage, as his increased
experience with his master's habits of hunting
cannot fail to make him understand more fully
the duties required of him; and certainly thus
securing in one dog the usefulness of two must
prove a decided gain. Liver or liver-and-white
is the best color for concealment; black or white
is too conspicuous, and may alarm the ducks.

You can hardly begin too early to teach your dog. He should, first, after learning his name, be taught to drop, whether by your side or at a distance, instantly, at the word or signal of hand, and to lie quietly until permitted to rise again. Do not allow him to rush out on the report of the gun to recover game, but make him wait until ordered to do so. Observe Markham's advice on the subject, a thorough sportsman and writer, whose book, "Hunger's Prevention," was published in the year 1655: "But by all meanes you must have your Dogge in such true obedience, that hee may not stirre from your heeles or let so much as his shaddow be perceived, till you have shot and yourselfe bid him goe, for to rush forth too suddenly or upon the first fire or clap of the Snaphaunce, though the piece goe not off (as many mad-headed currs will doe), is many times the loss of much good sport; which to avoyd suffer not your Dogge to stirre till you bid him."

It is frequently unnecessary to gather your dead birds as soon as killed, especially in still, shallow water, when it is better to allow them to remain until the shooting slackens or is nearly over, as the continual going and coming of the

dog will certainly alarm many ducks that otherwise might have "come in" fearlessly. Cripples,
however, should always be secured as soon as possible, and this an old, well-experienced dog will
generally do without waiting for orders; often
dashing out before the bird has struck the water,
knowing full as well as its master, from its manner of falling, whether the bird is dead or not,
and in such a case it would certainly be folly to
detain him. It is really wonderful how soon a
dog, if properly taught, will learn to understand
his duties, and it would almost seem to compre-
hend the reasons for them. I have seen old dogs
who were so fully up to their business that they
scarcely needed speaking to the whole day, taking
unordered a position where they would be screened
from sight of the ducks, remaining motionless
when they were approaching, and fetching cripples
as soon as possible, though leaving dead ducks
seemingly unnoticed until ordered to retrieve
them. Such a dog, I hardly need say, is inval-
uable, and never to be found, unless having had
great experience with a suitable and competent
master.

When teaching your dog to fetch, insist that
he shall deliver into your hand. If taught to

bring to your feet, when retrieving game he may release cripples where they will be enabled to escape, or at least cause unnecessary trouble to again secure.

Duck-dogs are usually rather hard-mouthed, being frequently obliged to grip tightly to prevent cripples from escaping; as ducks, I think, are more inclined than land fowl to struggle when captured, besides being considerably stronger and heavy to carry. My own dog, which I trained from a pup and made my almost constant companion, I taught to bring wounded birds tenderly in the following way:

One day, as he happened to be amusing himself gnawing splinters from the round of my chair, a kind of diversion he appeared to be particularly delighted with, I told him to quit, at the same time tapping him slightly on the nose with a small stick which I happened to have in my hand. He stopped, but presently commenced again. It immediately occurred to me that here was a splendid opportunity for teaching him the meaning of a command which it might be well for him to understand thereafter. So saying to him, " Don't bite it," I tapped him again. He of course stopped, not because I told him to, but to wonder at the cause of the rapping. On his beginning again, I would command and rap him as before, always giving the command (the same one every time) just before the rap, until, associating the command with the rap to come afterwards and his biting together, he soon learned to stop on hearing it in time to escape the punishment. I persevered in this way, sometimes giving him food and restraining him by this command from eating it, until at the end of two weeks (I had previously taught him to fetch) he would bring to me from a distance pieces of meat or bread without attempting to eat them unless permitted. The first crippled bird he ever retrieved (a wing-tipped pigeon) he carried over two hundred yards, delivering it into my hand without apparently hurting it in the least, and on his first day's experience with game he retrieved for me in the best manner thirty-seven ducks, mainly mallards.

Accustom your dog to retrieve from the water in the summer-time. If you commence to teach him in water too cold, he will learn to dread it; but, if a love for the sport be instilled into him before he learns to fear it, he will never refuse to retrieve, no matter what the temperature may be.

Frequently throw the object to be retrieved unobserved by him to a distance, or hide it in some easily accessible place; then encourage him to search for it, and, when needful, indicate its direction by a wave of the hand and arm. Throwing it over a fence or house, so that he may observe its direction, but not be able to see where it strikes, is also good practice. Let his lessons be short, yet frequent, remembering the more thoroughly these early lessons are learned, the more useful he will prove in after-life.

He should be taught to come to heel when ordered, and to remain there until permitted to go on, and should never be allowed to chase rabbits; otherwise when he is required for retrieving he may be having a little hunt on his own account.

Never whip your dog unless you are satisfied he understands what it is for, and let it be as soon after the committing of the fault as possible. Do not go at it in a merciless, inhuman manner, simply to vent your passion on the poor animal, who perhaps misunderstood your orders; and, instead of kicking or clubbing him, thereby possibly breaking a bone or otherwise seriously injuring him, use a whip or switch, which will sting sufficiently without bruising. Two or three strokes, rating him at the same time, are better than a dozen (or, as some do, "lick" till their arm is tired), for with too much punishment the dog forgets the fault in his desire to escape.

CHAPTER VI.

CAMPING OUT.

As a rule, the best shooting is not to be had near good hotel accommodations; consequently, if the sportsman would enjoy it, he is often compelled to sacrifice a few creature comforts, and be contented for a time with perhaps less desirable quarters. By many, however, possessed of hardy, vigorous constitutions and a keen love of the sport, this very opportunity for getting away from the trammels of society to the unreserve and freedom of the hunter's camp is often considered as even a greater enticement than the increased quantities of game.

I am satisfied but comparatively little is known by many sportsmen concerning the daily routine and business of camp-life, and much unnecessary labor and trouble is therefore undergone during their first experiences. A great deal of useless luggage is often taken, while much that should

be taken is overlooked or forgotten. I will, therefore, for the benefit of such of my readers as may
be in want of it, venture to offer a few plain
suggestions, trusting they may be turned to good
account.

Camping Out.

For from two to four persons and cook (full as
many as are desirable on a ducking expedition), a
regular house-shaped tent, about nine by twelve
feet on the ground, with a perpendicular wall of
about three feet in height, is as convenient as any.
This may be made of sail-duck as most durable,

but good heavy "drilling" will be found full as warm and impervious to rain, besides being much lighter for transportation, and, with proper care, it will last two or three years. A new tent should be well wet before using, or the first heavy shower may beat through, causing, perhaps, some inconvenience. As a further precaution against rain, a fly, as it is called, should be provided. This is simply a sheet of light cloth sufficiently large to completely cover the roof of the tent, which, however, it should not touch, excepting at the ridgepole, but should be drawn high enough at the eaves to leave a space of three or four inches between it and the tent. This breaks the force of the storm. Be careful not to touch the inside of the roof during a heavy rain-storm, especially if the tent has no fly, for the water will gather and run through wherever the tent may be touched, when otherwise it would run down outside.

Always pitch your tent on as high ground as convenient, on a little knoll, if possible, with the ground slanting slightly on all sides, so that water may not run into it. Without this precaution it is often necessary to dig a little ditch around the tent to conduct the water off; but, if the tent is pitched in a hollow or depression, even this is often of no avail, for as soon as the ditch is filled the water begins to come in, while cheerfulness and comfort depart. Never select a camping-place under large trees. They may fall down and perhaps kill some one, or cause other serious damage. Rather choose a situation amongst low, dense bushes or brush, which will shield the tent from heavy winds, and always pitch your tent with the open end towards the south. If near the river, see that the ground is sufficiently high to prevent inundation in case the water should rise.

Instead of cutting new poles and stakes whenever a change of camp is made, I would advise the providing of a permanent set, to be removed with the tent. They save a great deal of unnecessary labor, the tent sits better, and there is always plenty of room for them on boat expeditions.

Very few parties start out nowadays without the luxury of a camp-stove, and no one that I ever heard of, having tried it thoroughly, ever cared to give it up and go back to the old log-fire again, especially during cold or rainy weather. They may be made in a variety of ways, several of which I have tried, and the following plan I consider best answers my requirements:

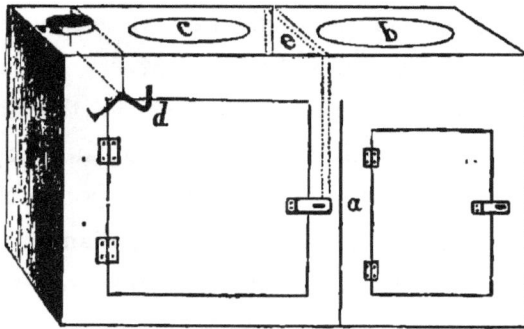

Camp-Stove.

Material, sheet-iron; length, 2 feet; breadth, 14 inches; height, 15 inches; oven, 10 inches square by width of stove, set in 2 inches in front of back end, and 3 inches below top of stove, thus leaving a flue of 2 inches deep underneath. At a, a partition formed of two thicknesses of heavy sheet-iron is riveted strongly to sides and bottom, $2\frac{1}{2}$ inches in front of oven and equal to it in height; in front of this is the fire-box. The kettle-holes are b, $7\frac{3}{4}$ inches, and c, $6\frac{3}{4}$ inches, in diameter; each strengthened around the edges by cast-iron rims or wiring, and provided with movable sheet-iron covers, which should be saucer-shaped, to prevent irregular warping. The damper is represented at d, and extends across the stove; by turning it either way the flames are made to travel over or under the oven, as may be desired.

A brace, *e*, of thin wrought-iron to strengthen the top of the stove and to prevent its warping, is firmly fastened on the inside across the top and part way down each side. Around the stope-pipe hole should be a rim of sheet-iron, upon which the pipe is fitted. Pipe, 3½ inches in diameter, made in tapering sections, which telescope together for transportation. When in use, the pipe should not be allowed to touch the tent, but a loose tin collar should surround it where it passes through, and be sewed in position to the cloth. The draught is regulated commonly by slide-gates, but various methods may be employed. Cost of stove complete, about $10 to $12. For cooking utensils: 1 frying-pan, 2 tin kettles, 1 coffee-pot, 2 sheet-iron baking-pans, 1 bread-pan, 1 dish-pan, and a large iron spoon are needed.

For a table, if you can procure a box such as is used for packing large plate-glass in, you will be suited to a nicety. Have the top planed smoothly, and set it up on legs to the proper height, then between the two sides of the box you may stow away your dishes when not in use. A tin cup, knife, fork, and spoon are, of course, needed for each person, besides a few extra plates and pans to use in cooking and upon which to serve food.

If you are out only for a few days or whilst travelling, it may be better to sleep on the ground than to go to the trouble of making a bed; but let me assure you, if you are located for any length of time, you will find a bed much more neat and comfortable. All you. need to carry besides your ordinary bed-clothes is a common bed-cord, and the labor of building the bed is almost inconsiderable. Four stakes, three to four inches thick and about four feet long, are cut from the neighboring trees, and driven firmly and to equal depth into the ground, in the angles of the parallelogram the bed is to occupy; six feet in length by four and a half in breadth, being proper for two ordinary persons. Two other strong poles are next cut the intended length of the bed, and fastened, one · on each side, about six inches below, and connecting the tops of the stakes. Each stake being strongly braced to prevent its springing in sidewise, the cord is wound tightly around and across, from one side pole to the other, the entire length of the bed, the turns of the line being about four inches apart. Other poles are now cut and fastened in position for side, head, and foot boards; the cording covered with a rubber blanket or spare quilt; leaves, weeds, flags, corn-husks, hay, or straw (as may be most convenient) piled on to the desired depth; the remaining quilts and blankets laid on smoothly, and, with the exception of the pillows, the bed is complete. These last may be composed of old boots, coats, empty boxes, powder-kegs, or possibly feathers, if the party has killed and picked game enough. Grape-vines may be used instead of the cord, if it cannot be easily procured, and, barring extra trouble in building, answer full as well.

Under the bed you may store your spare ammunition, clothes boxes or bags, and such sundry articles as are not needed for every-day use.

Two large chests should be made for ammunition, provisions, etc., and a third—or, what is quite as good, a stout waterproof bag—for clothes and sundries. The boxes should be made of pine, and of dimensions proportioned to the wants of the party.

The quantity of provisions to be taken will of course depend upon the number of persons to be provided for, and the intended duration of the trip; also, upon whether it will be convenient to procure more at any time upon the way or not. The habits and tastes of those who are to use them will, of course, determine their variety. Old hands at the business always learn to do without many luxuries, rather preferring hearty, nutritious food with hunger for their only sauce; and the more experience they have the less they are inclined to bother themselves with variety. The usual provisions carried by the market hunters are as follows: flour, corn-meal, pork, beans, coffee, sugar, salt, pepper, baking-powder, molasses, and onions, if procurable; to these, or such of them as he wishes, the novice may add what luxuries he may think proper—the fewer the better; and a good supply of matches, as well as soap, towels, gun-rags, and oil for lanterns must not be forgotten.

For tools, a saw, axe, and auger should be carried; a few nails, too, often come handy, and may be needed to mend a wrecked or leaky boat. A fish-line, with a few assorted hooks, might help to procure a change of diet, and should not be neglected. An old powder-keg, with the head taken out and fitted with a rope-handle, makes an admirable pail; and, if sawed in two, the bottom makes an excellent basin in which to wash the hands and face.

CHAPTER VII.

MISCELLANEOUS HINTS.

Clothing.—The color of the wild-fowler's dress should, as nearly as possible, be that of his natural surroundings, or, at least, be of some dull, neutral tint unlikely to attract particular attention; for, if it contrast too strongly with its background, any slight movement of the shooter will be likely to be instantly detected by his game, and his immediate locality afterwards carefully avoided.

In the fall, when the leaves and weeds are turning yellow, a light-brown or yellowish-drab will be found as good as any; whilst in spring, when the trees are more devoid of foliage, a suit of "pepper-and-salt" cloth is better, being less readily distinguished from an old log or stump. For timber or overflowed-prairie shooting, the "pepper-and-salt" is particularly recommended. Remember, too, ducks appear to apprehend danger more from the very dark colors than from the lighter ones.

Due regard should be paid to the temperature
of the climate and seasons. In August or early
September, a thin linen coat with large pockets
for game and ammunition is the proper thing;
while, as the weather grows colder, something
thicker must take its place. And in winter and
the colder fall and spring months flannels should
be worn next the body throughout. The "pepper-and-salt" cloth before mentioned for outside
wear will be found warm, durable, cheap, and
"for sale everywhere." For very cold weather
an English guernsey is one of the best things
possible. With a good coat outside, a person
can stand almost any needed exposure; and, being
flexible and easy, though close-fitting, this does not
interfere with free movements of the arms and
body.

Thick woollen gloves are preferable in cold weather to any others. If by accident they become
wet, wring them out dry as possible, and they
will be nearly as warm as before. In case
of necessity an old shot-sack drawn over the
hand will be found quite serviceable, especially
when picking up decoys or while paddling when
the handle of the paddle may be wet or covered with ice.

Do not let false pride induce you to buy close-fitting rubber-boots. They are intolerable in hot weather, and on very cold days you may wish to wear two pairs of stockings, which would then be impossible. Neither have them of too large a size, but choose a happy medium; for, if too large, they will soon crack across the wrinkles, and are then worthless.

A waterproof coat should always be carried in the boat in anticipation of rain, especially if hunting deep-water ducks, for during a shower they frequently fly much better than at any other time.

The Ammunition-Box.—When shooting from a boat with a breech-loader, an ammunition-box should be carried, to contain the cartridges and other ammunition, and to prevent their getting wet, as may frequently happen in a leaky boat or during rainy weather if no such provision is made. It should be made of wood or tin, waterproof, and large enough to hold cartridges sufficient for a good day's sport, with loading-tools and loose ammunition sufficient to refill the empty shells during the day if desired. It should be divided into two or three compartments for keeping separate the cartridges and other articles, and be provided with a hasp-lock, and leather strap on the side to serve as a handle for carrying. My own, of three-eighths inch pine, strongly dovetailed together, in which 1 usually carry 180 short 10-bore shells, loading-tools, three pounds of powder, half a sack of shot, and wads and caps

The Ammunition-Box.

to match, is 15 inches long, 12 inches wide, and 5 inches deep, and is divided into three compartments, as in the small foregoing sketch. The powder-canister is of tin, square, and made to fit in proper place. The loading-tools, all that are necessary with Sturtevant shells, are: a rod for pushing down wads, pressing on and ejecting caps; a tapered metallic tube through which (to insure them a level position and to prevent their edges from being injured) the wads are pushed into the shell; and a short piece of inch pine plank, bored partly through with sixteen holes, to receive the bases of the shells and to hold them erect for loading.

Oil for Locks, etc.—No vegetable oil should ever be used upon gun-locks; it is liable to gum, and thus interfere with their free working. Porpoise oil or refined sperm are the best for the purpose, and but very little is required. Porpoise oil is the kind generally used by gun-makers. To prevent rust, almost any kind of animal or fish oil free from suet is good; and for the stock, linseed-oil well rubbed in gives a nice polish, and will prevent water from penetrating. Caked dirt or a slight rust may be easily removed from the interior of a barrel by scouring with wet wood-ashes.

Powder.—In the West it is customary to use a much finer-grained powder for duck-shooting than is employed by the sea-coast shooters; but this I consider due more to habit than because any better results can be obtained from it; in fact, of two of the best duck-shooters of my acquaint-

ance, one uses F.F.G., very fine, and the other
No. 1, exceedingly coarse. Neither can be per-
suaded to use anything different, and both kill,
it seems to me, equally far. For my own part,
I am not so particular concerning the *size* of
grain, if it is only *uniform* and even; but where
I commence the season's shooting with a certain
size, I dislike to change to another. F.F.G. is
certainly fine enough for any one, and F.F.F.G.
unfit for use in a shot-gun.

For cleanness, strength, and evenness of grain,
no powder can, in my opinion, excel that manu-
factured by the Oriental Company, of Boston. I
have used many kegs of various kinds, and prefer this to all others. When I finished shooting,
last spring, having been using Oriental powder,
though my gun had not been cleaned for nearly
three weeks, and had been fired almost every
day, and on several days over one hundred times,
it was scarcely perceptibly foul, and might, to all
appearances, have been fired as many times more
without detriment to its shooting or inconvenience
in loading. The strength was equally extraordinary. My partner (who probably has killed as
many ducks as almost any man in the West)
used to remark almost daily: "This is the best

powder I ever shot, I really believe." These perhaps rather partial statements are in no wise in-
fluenced by any desire to favor one person or
firm more than another, but are given simply for
the benefit of such of my sporting friends as are
continually asking, " What powder do you use ? "

Minute exactness is often necessary in measuring rifle charges, and the Troy standard has therefore been adopted, while charges for shot-guns
not requiring this particular nicety are weighed
by the avoirdupois scale. This being the case,
many persons, not being acquainted with the rela-
tions which the two standards bear to each other,
are at a loss to properly compare them. To aid
such I append the following :

One pound avoirdupois contains 7,000 grains.

" " troy " 5,760 "

" ounce avoirdupois " 437½ "

" " troy " 480 "

" drachm avoirdupois " 27¹¹⁄₃₂ "

The pound, ounce, and grain, apothecary, are
the same as those of the troy standard.

Shot.—It would prove of great convenience to
sportsmen, especially in comparing the shooting of
different guns, if shot could be made of uniform
weights and sizes throughout the country ; and

with this object in view a committee were appointed by the New York State Sportsmen's Convention, held at Niagara Falls, to determine upon a suitable scale, and submit it for adoption to the various makers. This was done, and though their scale was accepted and the shot numbered accordingly by several leading manufacturers, yet the number of pellets contained in the ounce of their several makes was found to vary widely, which fact has given rise to much heated discussion. To me uniformity seems practically an impossibility; for lead taken from the same mine, and worked as nearly as may be in the same manner, will frequently be found to vary somewhat in density, owing to several reasons—it may be to a slight change in the degree of heat applied in some part of its reduction or refining, or, possibly, to some peculiar condition of the atmosphere. Then again, as the shot are sifted or assorted, the pellets which remain in a certain colander cannot be of precisely the same size, and consequently the numbers contained in separate ounces of the same manufacture may also, and do, vary. For instance, a No. 5 shot, which, according to the scale, should be $\frac{12}{100}$ of an inch in diameter, is, in fact, any shot sufficiently small to pass through a circular hole $\frac{13}{100}$ of an inch in diameter, and yet too large to pass through a similar $\frac{12}{100}$ hole. The following tables, showing the average number of pellets to the ounce of the various sizes, as now made by the leading manufacturers, may prove of utility to sportsmen ;

T. O. Leroy & Co., N. Y.			Tatham & Bros., N. Y.		
Dia. in inches.	Size.	Pellets to oz.	Dia. in inches.	Size.	Pellets to oz.
$\frac{21}{100}$	T T	32	$\frac{23}{100}$	F F	24
$\frac{20}{100}$	T	38	$\frac{22}{100}$	F	27
$\frac{19}{100}$	B B B	44	$\frac{21}{100}$	T T	31
$\frac{18}{100}$	B B	49	$\frac{20}{100}$	T	36
$\frac{17}{100}$	B	58	$\frac{19}{100}$	B B B	42
$\frac{16}{100}$	1	69	$\frac{18}{100}$	B B	50
$\frac{15}{100}$	2	82	$\frac{17}{100}$	B	59
$\frac{14}{100}$	3	98	$\frac{16}{100}$	1	71
$\frac{13}{100}$	4	121	$\frac{15}{100}$	2	86
$\frac{12}{100}$	5	149	$\frac{14}{100}$	3	106
$\frac{11}{100}$	6	209	$\frac{13}{100}$	4	132
$\frac{10}{100}$	7	278	$\frac{12}{100}$	5	168
$\frac{9}{100}$	8	375	$\frac{11}{100}$	6	218
$\frac{8}{100}$	9	560	$\frac{10}{100}$	7	291
$\frac{7}{100}$	10	822	$\frac{9}{100}$	8	399
$\frac{6}{100}$	11	982	$\frac{8}{100}$	9	568
$\frac{5}{100}$	12	1778	$\frac{7}{100}$	10	848
			$\frac{6}{100}$	11	1346
			$\frac{5}{100}$	12	2326

St. Louis Shot Tower Co.		Chicago Shot Tower Co.		
Size.	Pellets to oz.	Dia. in inches.	Size.	Pellets to oz.
000	33	$\frac{23}{100}$	0000	22
00	39	$\frac{22}{100}$	000	27
0	46	$\frac{21}{100}$	00	33
B B B	51	$\frac{20}{100}$	0	38
B B	60	$\frac{19}{100}$	B B B	46
B	71	$\frac{18}{100}$	B B	53
1	90	$\frac{17}{100}$	B	62
2	100	$\frac{16}{100}$	1	75
3	118	$\frac{15}{100}$	2	92
4	159	$\frac{14}{100}$	3	118
5	237	$\frac{13}{100}$	4	146
6	299	$\frac{12}{100}$	5	172
7	385	$\frac{11}{100}$	6	216
8	509	$\frac{10}{100}$	7	323
9	700	$\frac{9}{100}$	8	434
10	1103	$\frac{8}{100}$	9	596
		$\frac{7}{100}$	10	854
		$\frac{6}{100}$	11	14_4
		$\frac{5}{100}$	12	2400

WALKER & PARKER (ENGLISH SHOT).

Size.	Pellets to oz.
A A	40
A	50
B B	58
B	75
1	82
2	112
3	135
4	177
5	218
6	280
7	341
8	600
9	984
10	1726

MOULD SHOT.

L G	5½
M G	8½
S G	11
S S G	15
S S S G	17

Antidotes for Bite of Rattlesnake and Poisoning by Strychnine.—It is not an uncommon occurrence in certain sections of the Western country for dogs and even men to be bitten by rattlesnakes. The best remedy in such cases (which if neglected usually prove fatal) is whiskey, taken internally as soon as possible, and in sufficient quantity to produce intoxication, manifestations of which may be considered as certain evidence that the action of the venom is neutralized and the patient cured. For a dog, a pint will in most cases be needed, and a man may take fully a quart before showing signs of its effects, so powerful is the controlling influence to be first subdued in the venom.

For cases of poisoning by strychnine, liquefied lard should be given immediately in large quantities.

Pocket Compass, Wood-Craft, etc.—A pocket compass is a very essential article of the duck-shooter's outfit; without it, on dark, lowery days, he may easily lose his way in the tortuous and intricate windings amongst the tangled weeds and bushes of the swamp, and perhaps experience considerable trouble and uneasiness before finding it again. In the woods, a man who is versed in wood-craft has many things to guide him. The moss which grows upon the trees he knows is partial to the shade, and therefore always thickest on the north side; on the warm and sunny side of the tree (the south) he knows the branches are most frequently the largest, and his course is governed accordingly. In the swamp, he may be sometimes guided by the direction of the wind; but this is at best a fickle resource, for, should it change, it might lead him in a direction contrary to that desired. Perhaps the water may flow slowly through the swamp, and this may be sufficient to guide him; but it is oftener stagnant, and then affords no clue. To be on the safe side, carry a compass.

You may determine from which side the wind comes, even when the air is seemingly still, by holding above your head your wet finger, which you have previously held in your mouth until warm; it will be plainly felt to cool first on the windward side.

Cracked Hands.—Duck-shooters are frequently troubled with chapped and cracked hands; the alkaline deposits of burned powder, and the continual wetting of the hands whilst picking up game and decoys, rendering such a condition in cold weather, without constant and particular care, almost inevitable. As a precautionary remedy, diluted vinegar is, I think, as good as any ; for by its peculiar acid action the harmful properties of the alkali are neutralized. The skin should also be kept soft and pliable by frequent applications of glycerine, camphor-ice, or tallow, particularly just before going to bed.

Rheumatism.—The wild-fowler is often liable to get wet from being caught without proper protection in a heavy rain-storm or otherwise, when acute rheumatism or a severe cold may follow, in anticipation of which, and to prevent such undesirable consequences, liquor, such as whiskey or brandy, should be carried and taken internally when necessary, but should never be resorted to at any other time. I am sorry to record it, but it is a fact, that many parties start out from home on a camping expedition with the ostensible purpose of shooting ducks, when in reality an opportunity for unrestrained whiskey-drinking is their main object. I once happened to call at a camp where four fellows were "roughing it" for a few days. Seeing no one outside the tent, I ventured to look in. Lying on a lot of straw, which was scattered about the floor fully a foot in depth, were all four, drunk and sleeping soundly. By some of their restless, unconscious motions the straw had been pushed against the hot stove, and when 1 looked in had already commenced blazing. 1 of course extinguished it as soon as possible, and awoke them. Had I not providentially chanced to call as I did, some one of the party, if not all, would in all probability have been fatally burned, for all were too stupefied to save themselves.

As a local remedy for acute muscular rheumatism, a mustard-plaster placed immediately upon the part affected will, in most cases, soon prove effectual. If the white of an egg be mixed with the plaster, no blistering will ensue.

Ague.—Never start out to shoot in the morning with an empty stomach. Eat something, if it be no more than a cracker; you will be less liable to be attacked by ague. Be careful, also, to avoid drinking warm slough water. Different remedies are required by different persons to cure ague;. as for myself, I have so far got along without requiring any, never having experienced a touch of " the shakes."

CHAPTER VIII.

The Mallard (Anas boschas).—Adult Male: Bill about the length of the head, higher than broad at the base, depressed and widened toward the end, rounded at the tip. Upper mandible with the dorsal outline sloping and a little concave; the ridge at the base broad and flat, toward the end broadly convex, as are the sides; the edges soft and rather obtuse; the marginal lamellæ transverse, fifty on each; the ungius oval, curved, abrupt at the end. Nasal groove elliptical, sub-basal, filled by the soft membrane of the bill; nostrils sub-basal, placed near the ridge, longitudinal, elliptical, pervious. Lower mandible slightly curved upward, with the angle very long, narrow, and rather pointed, the lamellæ about sixty.

Head of moderate size, oblong, compressed. Neck rather long and slender. Body full, depressed. Feet short, stout, placed a little behind

the centre of the body. Legs bare a little above
the joint; tarsus short, a little compressed, anteriorly with scutella, laterally and behind with
small reticulated scales. Hind toe extremely
small, with a very narrow membrane; third toe
longest, fourth a little shorter, but longer
than the second; all the toes connected by reticulated membranes, the outer with a thick margin, the inner with the margin extended into a
slightly-lobed web. Claws small, arched, compressed, rather acute; that of the middle
toe much larger, with a dilated, thin, inner
edge.

Plumage dense, soft, elastic; of the head and
neck, short, blended, and splendent; of the other
parts in general, broad and rounded. Wings of
moderate length, acute; primaries narrow and
tapering; the second longest, the first very little
shorter: secondaries broad, curved inward, the
inner elongated and tapering. Tail short, much
rounded, of sixteen acute feathers, of which the
four central are recurved.

Bill greenish-yellow. Iris dark-brown. Feet
orange-red. Head and upper part of neck deep
green, a ring of white about the middle of the
neck; lower part anteriorly, and fore-part of

breast, dark brownish-chestnut; fore-part of back light yellowish-brown, tinged with gray; the rest of the back brownish-black; the rump black, splendent, with green and purplish-blue reflections, as are the recurved tail feathers. Upper surface of wings grayish-brown; the scapulars lighter, except their inner webs, and with the anterior dorsal feathers minutely undulated with brown. The speculum, or beauty spot, on about ten of the secondaries, is of brilliant changing purple and green, edged with velvet black and white, the anterior bands of black and white being on the secondary coverts. Breasts, sides, and abdomen very pale gray, minutely undulated with darker; lower tail coverts black, with blue reflections.

Length to end of tail, 24 inches; to the end of the claws, 23; to the tips of the wings, 22; extent of wings, 36; wing from flexure, $10\frac{1}{2}$; tail, $4\frac{1}{4}$; bill, $2\frac{2}{12}$; tarsus, $1\frac{3}{4}$; middle toe, $2\frac{2}{12}$; its claw, $\frac{5}{12}$; weight from $2\frac{1}{2}$ to 3 lbs.

Adult Female: Bill black in the middle, dull orange at the extremities and along the edges. Iris as in the male, as are the feet. The general color of the upper parts is pale yellowish-brown, streaked and spotted with dusky-brown. The feathers of the head are narrowly streaked;

of the back with the margin and a central streak
yellowish-brown ; the rest of the scapulars similar,
but with the light streak on the outer web.
The wings are nearly as in the male ; the speculum similar, but with less green. The lower
parts are dull olive, deeper on the lower neck,
and spotted with brown.

Length, 22 inches ; weight, from 2 lbs. to 2½.

The young aequire the full plumage in the
course of the first winter.

Mallards breed in small numbers in the various
swamps and sloughs of the Western country, but
by far the greater number betake themselves to
the unknown regions of the north, and there,
unmolested, rear their young. The month of
August is hardly over before they again begin to
make their appearance in the more northern of
the Western States, but the shooting of them
cannot be said to have fairly commenced until
about the middle of September. After that time,
and until the freezing of the waters drives them
further south, the numbers killed are sometimes
almost incredible. Their habits vary considerably in the different localities which they frequent, and at different times, and various means
are employed for their capture, the most practi-

cal of which it is my intention to explain as clearly as possible.

For convenience' sake, and to ensure a more thorough description of details, I shall, in the remainder of this chapter and that immediately following, adopt the conversational style, addressing my remarks to a supposed novice, who is about to take his first lessons in duck-shooting, the present time being the evening preceding the sport; place, hotel near the shooting-grounds.

We must start early in the morning; so get your gun and ammunition ready, and don't be sparing of the latter, for it is much better to have to bring some back than to leave good shooting for want of it. Sometimes, when least expected, a person will find all the ducks he can reasonably wish for. Chicago sixes or St. Louis fives are about the shot you need, as at this season the ducks are not very full feathered, and the mallard is not over-tenacious.

Load your shells with four drachms of powder. " Rather a large charge of powder," you say. You perhaps have been used to shooting quail or woodcock, where smaller charges are sufficient, the shooting always being close.

An ounce of shot is enough; there is no need of

more. The gun will kick harder and shoot weaker; besides, if you add half an ounce to each of those cartridges, you will have considerable extra weight to carry. You can kill as far with the ounce if you hold right. As we shall probably have to leave the boat, take along your sack for carrying shells; you might carry them in your pockets, but the sack will be handy to hang on a branch or to lay down while shooting, and it will be much more comfortable than having the weight in your pockets. You will need your long rubber boots, as we shall probably have some wading to do. But put a pair of slippers in your pockets; you can wear them while in the boat, and will find them much easier than the boots this warm weather.

Breakfast will be ready before daylight, and I will see to the luncheon; for if we find the shooting decent, we shall most likely stay till dark.

I am glad to see you wear a hat to shoot in. A cap I abominate. If it rains, the water is continually dripping and running down the back of your neck; and when the sun shines fierce and hot, it furnishes no shade, as does the hat. Very often, too, as a person is taking sight, the rays of the sun strike one's eyes, dazzling them, and thereby causing a miss. Well, good-night; I will call you in time in the morning.

*   *   *   *   *   *

Why, here it is after four o'clock. I came very near oversleeping. I must call my young friend; but here he is, coming to call me! Good-morning. You are on hand, sure enough. Suppose you feel like giving the ducks particular fits to-day. Well, we'll see what we can do for them very soon. The sky promises fair weather, and we have a cool west wind to refresh us while rowing to the shooting-grounds, and which, if it continues, as 1 think it will, will make the ducks feel more like moving about than they might be inclined to do if it were warmer. Breakfast is ready, I hear from the dining-room; so let's go in and see if we can do it justice. There's not a big variety, to be sure, but it's good stuff to last and to work on, this corn-bread and duck; and it isn't very apt to .produce dyspepsia, especially if one does much rowing, and tramps far in the muddy bottoms for exercise. It just suits you, does it? Well, eat heartily, for you may have work to do, and, as Joe Carroll used to say, "A man that can't eat isn't fit to do much of anything else."

Keep on eating. 1 am only going out to call Jack, the dog, to his breakfast. Here he is; what do you think of him? A pure-bred setter, you see, and I think as good a retriever as can be found. He also understands a thing or two about quail and chicken shooting, as 1 may have an opportunity of showing you before long. Sportsmen have peculiar fancies regarding retrievers, and among writers one advocates one variety, another another, and a third again perhaps a cross between the two; in my opinion, the main requisite, second only to power of endurance, is simply this: that the dog should take an especial *delight* in retrieving. No man can excel in any pursuit unless he has a particular liking for it, no matter how well adapted he may be. So it is with the dog; no matter what particular breed he may belong to, if he has no actual love for the sport, no amount of breaking will make a decent retriever of him. 1 know of a dog to-day, a cross between a setter and a pointer. He is rather old now, but four or five years ago a better retriever for Western duck-shooting was not made. To look at. him, you would laugh at the idea of his being a duck dog. He looked more like a fighting dog, and, in fact, next to retrieving, fighting was his favorite pastime; and, excepting for this one purpose, he did not appear desirous of courting the society of any of his kind. He would weigh perhaps fifty pounds, smooth haired, very strongly built, and stood about twenty inches high. It was almost impossible to tire him out. I have known him to retrieve mallards all day in running water for four persons, quite constant shooting, bringing in over two hundred ducks, and going for the last ones as readily as for the first. He returned with a duck as quickly as possible; never walked or loitered on the way. He could tell when a duck was struck as well as the shooter, and would watch it as eagerly to see if it fell, when he would immediately go after it, sometimes five or six hundred yards.

I well remember on one occasion I was sailing down the Illinois River with a hunting party, and passing by "Clear Lake," a beautiful sheet of water, saw quite a large flock of mallards feeding near the shore of the lake. I seized my gun, and calling to the dog jumped into a paddle-boat, paddled ashore, and proceeded to "bushwhack" them. The grass was quite high, and by creeping low down on all fours I was enabled to get quite near them. The dog followed close behind me, crouch-

ing low and watching my every motion. Getting
as near as I wished, I fired both barrels, and succeeded in stopping thirteen, eight being dead and
five splashing round with wings broken or otherwise crippled. In went the dog. I ran down to
the water's edge, when he brought me out four
of the cripples, passing by the dead ones going
and returning, and not paying any attention to
them (the fifth skulked low down in the water,
and he did not see her). When he had delivered the fourth, he started along the lake
shore as fast as he could go. I called and
whistled; he paid no attention to me, but kept
on some hundred and fifty yards or so, and, diving into the grass, appeared with another crippled
duck, which he immediately brought to me. I
had no idea what he could be going after, but
his quick eye had seen the duck fall from the
flock when mine were engaged in another direction. After his retrieving the dead ones, we returned to the party with ducks enough for supper. I have often had him bring me over a
dozen crippled ducks in a day, while going to and
coming from the ponds—ducks that I had no idea
were near me. He would come across their trail
when they had run up into the woods from the

water, and then follow them up until he found them hiding in some old brush-pile or under a log.

But let us start. It is only a short distance to the river. Just take your gun and the lunch-basket, and I will carry mine and the ammunition.

There, this is our boat; put down the dunnage, and we'll launch it. Jack, you see, is in his place in the bow as soon as it is in the water. You may sit in the stern. Keep your gun handy, and 1 will row; it is not far. What a splendid river this is for boating, isn't it? Straight stretches for miles, and but very little current; the shores, you see, are quite bare and devoid of weeds, and offer little inducement for the ducks to light along them. A stranger who was not well acquainted with the habits of ducks would little think from the few he might see along the river what multitudes abound in this country. There! do you see that flock of ducks to the right, away beyond those tall trees? They are mallards, and are now over a favorite feeding-ground of theirs, called by the local hunters here the duck or rice pond. It contains perhaps three hundred acres, water from ten to twenty inches, and mud ten to twenty feet in depth; almost the entire surface is covered with the dense growth of the wild oats or rice, whose stalks grow to the height of ten to fifteen feet. The seeds of this cane are the favorite food of mallards and other shoal-water ducks in the fall, and when it grows to any great extent the ducks are usually quite numerous. This pond is fed chiefly by springs, and has for its outlet a small, crooked, shallow stream, call Mud Creek, which empties into another, known as Crow Creek, the mouth of which is only a short distance from here, just where you see that opening in the willows on the east side of the river, perhaps a quarter of a mile ahead. We will strike in there, and try a couple of hours' shooting in the rice-pond, until the morning flight is over, when we will go to another place I think rather favorably of for mid-day shooting. I will explain to you when we get there why the shooting at that time of day is better there than in the rice-pond.

Here we are at the mouth of the creek. It is not wide enough to admit of rowing, so you may take an oar and stand in the bow and paddle, or push against the logs or bank, as you have opportunity. I will guide the boat with this long paddle. Come forward to this seat, and sit perfectly still while I pass by you. I will look out for the balance of the boat, so don't you mind anything about it. There, that's good! Two nervous persons in a boat like this are very apt to get spilled out if they attempt to pass each other. Neither will trust the other; so in passing, if the boat cants either way, both throw their weights to the opposite side, and, instead of preventing the accident, it is thereby made certain. So remember this: let the one who passes keep the balance. Hark! just listen to the cracking of the guns in the rice-pond. The poor ducks are catching it now.

The moon is now in its first quarter, and the nights being quite dark the ducks remain later in the morning; but during the full of the moon, when they can see better, they feed nearly all night, and are ready to take their departure out of danger from the hunters much earlier. I have been in places where they were hunted a great deal, and during the full of the moon could hardly get a shot in the rice-ponds until after dusk, when they would come in by hundreds, and at daylight would leave as suddenly. After a very dark night they would seem more anxious to stay a while in the morning, and would stand considerable banging.

That fork to the right is the noted stream "Mud Creek." You may put down your oar now, as the current is not quite as rapid as this we're leaving, and get your gun ready; for in a few rods the creek widens, and we then come to the edge of the wild rice, which grows on either side of the creek, until we come to the pond. There are always a good many ducks sitting in the edge of the rice along the creek, and by moving quietly we can get very close to them before they take alarm. Sit with the left side rather more in advance, so that you can shoot ducks crossing to the right more easily; and don't shoot if you have to drop them too far in the rice, for it will take more time to break down rice and hunt them up than they are worth. Wait until you get one over the creek or close to the edge. We must not waste too much time here, you know, for it may be more valuable somewhere else. Now, no more loud talk for a few moments, until we see if we can't get a shot; and remember this last caution: Be cool; and don't shoot until you get your gun just right, for nothing but dead ducks count here —cripples are of little use.

Steady! Well! well! Why didn't you shoot?

Gun wasn't cocked, eh? Why, I've heard you cock it fully a dozen times since we struck the edge of the rice. I guess you must be getting a trifle excited. Well, it's to be expected at first, with so many ducks continually jumping up before you; but it will soon wear off. The trouble is, when you took the gun down after sighting at that wood-duck a moment ago, you let down the hammers, and forgot to raise them again. Look out next time, and keep cool; you'll have lots of chances.

Careful! Well done! That duck was neatly killed. No one could do better. A young mallard drake! Waited a little too long, didn't he? Pick him up as we pass. Always pick up ducks by the bills or heads, and shake them well before putting in the boat; their feathers hold a lot of water, and they look much better and will keep longer when dry than after lying in the wet all day. Lay him on his back in the bow, in front of the dog. I like to keep my ducks' feathers smooth too, not turned "every which way." Look sharp, now; in this bend ahead there are ducks, I'll warrant. Steady! I thought so! Well, I guess you a'n't much of a "slouch" at shooting, if this is the way you are in the

habit of doing. Two dead that shot! I expected
to see you kill the other with the second barrel. She flew straight up the creek. Why didn't
you try it? The surprise of killing both the
others with the first barrel took you off your
guard, and you didn't think of it, I suppose.
But you're doing well enough, any way. It's
easier to sit here and tell how than to take
your place and do it any better.

Do you see that low muskrat house there in
the edge of the rice? Well, remember where it
is, and when we come out look out for a shot at
a wood-duck there. You will be almost sure to
find one. They are very fond of sitting on such
places to preen themselves and bask in the sun,
and I have jumped many a one from the same
place. They are feeding now, and it's only from
about nine o'clock in the morning to three in
the afternoon that they frequent such places.

Mark! Do you see that flock of teal just in
the edge of the rice near those lily-pads?
Don't make a motion, and I'll see if I can't
steal on them behind that point of rice. Now,
watch sharp as we come into view; if they will
wait, give them the first barrel on the water,
and don't forget the second one this time.

There they go! (Bang! bang!) No, several have suddenly concluded not to go, 1 should judge; take my gun and shoot those cripples, if you can. That's the way to do it; seven more to help the pile. A'n't they pretty little fellows? And fat, too, as butter. Most delicious eating, as you'll find to-morrow.

Here we are at the pond. You can't see much of it, though, this rice is so high; but we will land as we come out on this point to our left, and then by climbing one of those tall trees you can have a better view of it. If the pond was new to me, that would be the first thing I should do. 1 could then see where the ducks were working most, and where the thin, open patches were in the rice, and, by taking my bearings properly, could go directly to them. My back path, too, would be the most direct way out. A person who didn't think of this might push around for hours without finding many ducks, and when he did stop he could not be sure of being in the best place; then if he could see nothing to determine his direct course out (which he would have to break anew), he would be compelled to follow his perhaps long and circuitous route back. Now, put these little hints "to soak," as they say out here; your success hereafter depends chiefly upon them.

We probably won't scare out any very large batches of ducks, as some of the other hunters have come in this way before us; but I think we will have some pretty good fun for a while, from the way the other boys are shooting. What a rattling they keep up! And just look at the poor mallards trying to find a place to light. They do hate to leave. Well, we'll take this path where some one has broken a road for us. I know where it leads to, and I guess we shall find a few ducks there. What a squawking they do keep up everywhere! Now, you see, this mudstick or setting-pole,* as we call it, which I have exchanged the oar for, comes into use? Without it we could hardly get along, or at most very slowly; for the mud is so soft you can push a paddle almost its length into it, and it is often harder to pull out than it was to push in. By putting this on the roots of the wild rice you have something solid to push against, and it does not enter the mud deep enough to stick much. The boat makes such a

* Pole with a forked or widened end to prevent its sinking in mud. See cit in the chapter on "Boats."

noise going through the rice you need not ex-
pect to shoot much until we get on a stand; the
ducks hear us too plainly.

Look! what a lot are getting out ahead of us.
That is where I want to stop. There is an open
place there in the rice for a few rods, and the
rice is rather lower and more thinly scattered
about it; there is a nice thick bunch in the
middle, too, if no one has broken it down, where
we can hide completely. Now you can see it
straight ahead. We will run the boat right into
the middle of it; there, I guess that will do well
enough. Take the paddle, and bend the tops of
the rice down over the bow, so as to hide the
boat a little better, and it won't be in the way
of the gun when you are shooting.

Ha! ha! don't be in a hurry to stoop; those
ducks are a quarter of a mile off, and no
more apt to come here than to go somewhere
else. You never need stoop until they get nearer
than that. How angry it has made me to have
a · nervous know-nothing catch me by the arm and
yank me down, for fear a duck that he happened
to catch sight of half a mile off would see me
and take alarm; a duck too, perhaps, that I had
been watching myself for two or three minutes.

"Greenhorns" and "hoosiers," as the regular hunters call such fellows, when they are hunting three or four together (and often half a dozen shoot from the same boat, usually a big fishing-skiff), always commence to cry, "Down! down! Here comes a duck!" whenever they see one any where, no matter how far off, or in what direction he may be going; then all crouch low in all kinds of positions. "Where? where?" the others cry, and the one who said "down" now says, "This way"; then the enquirers look at him to see which way he is looking, and commence turning round, getting partly up, and stretching their necks to discover the duck, when the discoverer says, "Oh! he's going; he a'n't coming this way." Then all rise up again and breathe freer, until another starts the "Down, down" again, when the same performance is repeated. When a duck is coming, he usually sees them while making so many motions, and of course sheers off; but sometimes when they are all standing up, watching, some young fool of a drake will come right up to them, unperceived, perhaps, until he is passing by. "Down! down!" again; and while the others are stooping, one fires, but not until the duck is out of range. Then they commence,

"What made you shoot?" "Why didn't you let
him go by?" "We could have called him back!"
And one perhaps is now trying to do so, making a noise more like a bull-frog than a duck,
and keeping it up until the duck is out of sight.
I am not over-drawing it a particle, as I have
often seen it.

Mark! to the right. I'll see if I can't call that
young drake this way. Yes, here he comes; don't
move until he gets almost to you, and then put
it on him without getting up, and pull. (Bang!
Bang!) You shot too quick, but I've saved him.
Remember this isn't like shooting along the creek;
there the ducks, jumping up in front and flying
most always directly away nearly on a level with
you, may be often killed with snap-shots, like
quail or snipe; but in this cross and overhead
shooting, overhead most especially, snap-shooting
won't do. You can't make the proper and necessary calculations unless you take more time,
and the position in overhead shooting is one you
are not much accustomed to in the field. Now,
the next one that comes, take it slower; if coming
in line, let him get an angle of forty-five degrees
or more with you before you raise the gun; then
bring it up directly behind him, moving it con-

siderably faster than he is coming, and timing it as
near as possible, so as to be just passing in front
of him as he gets almost directly over your
head; then press the trigger smartly (don't pull
it) without stopping the gun, and if he is inside
of thirty yards you will kill him clean; beyond
that distance you will need to hold a little
further ahead, according to the increase of distance.

Mark! straight ahead; here comes a flock directly at us. Don't be in a hurry, and pick out
your bird; take one of the hind ones. Good!
that fellow won't get away. I got a couple down
dead that time—that makes four altogether. We
must keep account of them, and mark where they
fall. We will pick them up when we get through;
three are in open sight, and one is in that little.
bunch to the right, remember.

Mark! Teal coming. Cool now, but not too
slow. There, that's business-like. We can't count
that wing a broken one, for he'll get away; the
dog might get him, but I don't want him clam-
bering into the boat all dripping with mud
and water, and there is no place for him to
stand on out there, so we will let that one go.
Mark! east. If they come, take the head one; I

will try and get the hind one. Good! Load up again; here come some more to the right. How quick they turn when they hear the call? Steady! Four down that time; those make fifteen, including the five teal.

See what a lot some fellow is putting up over east; we will be apt to get a crack at some of those. Yes, here they come. I'll leave the head ones to ·you. Only two down that time, eh? I missed that second shot of mine clean. Load up quick; we'll get several sho s at these before they all get settled. Did you see those two drop from that flock over there to our right? Some fellow is doing pretty good shooting there. I've seen him kill several lately. Steady! Low down to the left. (Bang!) He's all right; but never mind, no man can kill every time.

See that flock towards the sun coming into the pond; how quickly they lower when they get over it? There! that same fellow at our right just dropped another. That flock is coming straight towards us, but is rather high yet. Keep low, and let them circle; I'll call them right down. Did you see how they darted down when they saw those dead ones? They'll turn right back. Now, steady! That's the kind of fellows

we want.   Three more!   Every one makes the bag
bigger.   We are doing what I call mighty good
shooting; of course we are bound to miss some.
Yes, I see him; you can take care of him.
Well, he did let go everything, and all at once
too, didn't he?   " Dead as a stone !"

So the sport continues, until the ducks, as
though tired of performing their part of it, become fewer and further between; then the dead
ones are picked up, and we start again towards
the river, killing a few more while paddling
down the creek, the wood-duck, as I predicted,
contributing to the number.   The continuation of
the day's shooting I will describe in the next
chapter.   .

CHAPTER IX.

WE have done very well so far this morning—fifty-three ducks, I believe you said. Now, if you are not tired—and you surely don't look to be—we will try the place I mentioned. It is now about ten o'clock, and we can go there in half an hour, so I think we will have a look at it any way. We leave the boat in the bend just ahead of us, and then we have about a quarter of a mile to walk back into the woods. You can be putting on your rubber boots.

There, jump out, and take the guns and dunnage. Lay them down there on the bank until I moor the boat; it's always best when convenient, for there may be cattle here in the woods, and I have known them to step into boats, when pulled up on the bank, and break through the bottoms of them; our ducks, too, will be safer from hogs. I never carry an anchor or weight, but simply push the paddle down into the mud as far as possible, and fasten the chain to it. I guess that will do. Take your gun and ammunition. We will take the lunch with us, too; and there are a couple of bottles of lager-beer, which is much better to drink than slough-water. There, I'll take the rest. How much better walking it is in this bottom-timber than in the woods of New England; no underbrush to bother you, and the ground level and free of stones.

Do you see those large bunches of sticks and brush in the tops of those tall trees? What do you suppose they are? "Nests of some kind." Yes, they are the nests of the blue heron; there are hundreds of them here, you see. They come here every year about the middle of April, and commence to build their nests for rearing their young. They keep up a constant noise day and night during their stay, and can be heard a considerable distance. Here comes Jack with a crippled duck in his mouth. Well done, old fellow! Back, now! I want to have a look at those ducks ahead of us before you scare them off.

The slough is just there in front of us where you see that opening through the trees. Be careful, and don't step on a dry stick, and keep that large maple between you and the slough. See that flock come in, and what a quacking the

others are making over it! There, just look by
the tree, and see what a lot of them there are.
Down, Jack! What comfort they appear to be
taking; it almost seems a pity to disturb them.
Why, the slough is almost full. I suppose you
have picked out the thickest cluster, and are im-
patient to shoot. But no, that won't do; you
might perhaps get ten or a dozen with both barrels, but you would very likely spoil our shooting
for the rest of the day. We mustn't tell this
big body of ducks that we have guns here, or
they won't be very likely to come back; but I'll
just step out and put them up without alarming
them too much, and then they will return again
shortly, a few at a time; and, from the large number here now, I judge we will be kept pretty
busy taking care of them. If it was later in the
day, it might be to our advantage to try a sit-
ting-shot, as there might then be enough come back
to last us the rest of the day, and we would have
those killed at the first fire for decoys; but just
now I think we had better wait.

Halloo! What a flapping and quacking! You
go round to the right, and wade out, most to
the edge of the open water. I will go round
the other way, and take a stand opposite to

you. The ducks will come in between us, and
as the slough is not over fifty yards wide, we
will have some pretty shots. See them coming
back already! Just shoot away now whenever
you feel like it, till we get our stands fixed.
One down, but a cripple; you needn't shoot him
over. Jack will get every cripple that falls here,
I'll warrant you. Hang up your cartridge-bag on
a branch of the buck-brush, and cut some brush
to make your blind thicker. I can see you
through it quite plainly; and trim off the tops
of the brush in front of you about breast-high,
so that you can shoot over it more easily.

Mark! east. Let them go over once while I
call them. Here they come again, right low
down. Be careful, now, and don't shoot me; take
the foremost. Hurrah! Five down. Go on, Jack,
and fetch that cripple. Good dog! The ducks
almost always enter this pond from the east, as
the trees are much lower on that side; so when
I cry "mark," look to that end, unless I tell you
some other direction. Mark again! Well done!
I guess I will go out now, and stick up these
ducks we have killed for decoys; they will help
to quiet any suspicions in the new-comers. These
dead ducks make the best of decoys—far better

than any artificial ones. Cut me three or four
more sticks ; make them about two feet long,
and sharpen both ends. Come, Jack, old fellow,
and fetch me those dead ducks. There ! see how
naturally that duck sits on the water. You see
I push these stakes down firmly into the mud,
leaving about six or seven inches above water ;
then right between the base of the bill and neck,
on the under side of the head of the duck, there
is a soft place free from bone, and by pushing
this down firmly upon the point of the stick
the duck is held in position. The tail feathers .
usually need raising a little, as in life the mallard carries them slightly elevated. Many use a
crotched stake instead of one sharpened at the top,
and merely hook the duck's head over it. This
plan is full as good in calm water, but if there
happens to be a little swash the heads work
loose and fall down. There, that makes quite a
respectable-looking flock. Come, Jack, let's get
back to our blind now, the sooner the better.
In some parts of the country, where it may be
difficult to procure stakes, cane or rice stalks
may be used to stick up ducks on. But as these
are very brittle and elastic, they cannot be easily
pushed into the under part of the head; so

they should be cut considerably longer than sticks needed for the same depth of water, and one end being pushed down the throat of the duck, it should then be bent sharply just back of the head, and the other end pushed firmly into the mud.

Mark! Let them go over. I'll call them down. Coolly, now; and always take the head ones when coming in from that direction. Glorious! Every one clean killed. No work for you this time, Jack. By Jove, this kind of sport can't be beat anywhere. They drop their wings and come in so unsuspiciously right in between us here, and they can't get out either, except by "climbing" almost straight up; and as their headway is almost stopped, we can take our time with them. We ought not to miss a shot, but of course some unaccountably bad ones will be made. I remember doing some most miserable shooting here once, and my partner for the day, usually a better shot than common, doing the same as I. I can't explain it to this day, and I sha'n't try. Mark! Let them go by. Take it cool now, and see how "clean" you can do it. I'll trust you with both of them this time. Steady! Ah! that won't do; you were in too much of a

hurry with your first bird, for fear the second
one would get away. There was no need of that,
for after you had fired both barrels he was
not forty yards from you. Take plenty of time
with your first barrel; let the second be the
snap-shot, if either—but neither, if you can help
it. I see, though, to-day you are not one of the
kind to get discouraged because of an occasional
miss. Why, on the day when I did my poor
shooting here, I was standing in this very blind
and eating my lunch—my partner, whom I will
call Ned, was in your position—when a single
mallard duck came into the slough, low down,
intending to light within thirty feet of me. I
jumped up, rather too quickly, perhaps, and fired;
not a feather was touched, as I could see. Then
the second barrel more carefully, as I thought,
but no game. Ned laughed rather immoderately,
it seemed to me, as the duck flew directly towards
him, and, taking aim very deliberately, he fired;
the duck certainly was not ten yards away, and I
expected to see it literally blown to pieces; but
no, it simply changed the direction of its course.
I looked at Ned, who appeared a trifle more serious, and was taking aim again with a determined,
bloodthirsty expression of countenance. Bang went

the gun a second time, and the duck flew away, quacking wildly, scared most excessively, but apparently otherwise unhurt. It was now my turn to laugh, and I think I did so. Soon Ned had to join in. Well, we both agreed that such a completely scared duck we had never before seen. It actually didn't know which way to go; there was plenty of time for either of us to have fired a couple more barrels at it before it was out of range, and we could hear it quacking as though for a bet and against "big odds" some time after it had got out of sight.

Mark! By Jove! I reckon "somebody's cut the bag open," as the saying is out here, from the way they are coming. Leave the hind ones to my care. That's the kind; load again as soon as possible. Well, I'm ready. 'Way you go! Two more. Now pitch it into them coolly. This is exciting, but we must keep steady. Let this large flock light. It isn't exactly sportsmanlike, I think myself, but it is often excused in duck-shooting. When you can get three or four in range, blaze away; don't wait for me. Quiet, now; don't move. Bang, bang, ba—bang! There, Jack, is work for you. Four, five, six, seven dead and three cripples, and only four escape. This is slaughter.

Watch this old fool of a duck coming, and see me
"raise her." Quack! Quiet, Jack. Bang! What
a cloud of feathers! Fetch her, Jack. Well, well.
As Joe Carroll used to say, you'd think she'd
been shot with a threshing-machine; she feels like
a big sponge. 'Twill make good hash for some
Chicago boarding-house. Did you notice when I
called her, how quick she dropped her wings? I
like "calling by mouth" much better than with
a "squawker," especially if the ducks are passing
reasonably close. I will try and explain to you,
though, how to make a squawker, if the ducks will
only keep away long enough.

First a tube of wood or metal (bamboo cane
is chiefly used) is to be provided, about three-
quarters of an inch inside diameter, and from
four to eight inches long; a plug about three
inches long is fitted to one end, and after being
split in two, one-half is grooved to within a
quarter of an inch of its smaller end, the groove
being perhaps a quarter of an inch wide and
of the same depth. The tongue is simply a very
thin piece of sheet copper or brass, which should
be hammered to increase its elasticity; it should
be about two and a half inches long and from
three-eighths to half an inch wide. At one end,

which should also be thinner than the other,
the corners should be rounded.   The tongue is
then placed over the grooved half, the round
end nearly to the extreme smaller end of the•
plug, and the tongue completely covering the
groove.   The other half of the plug should be
shortened about an inch and a half from its
smaller end, and then being placed on the grooved
part, thus holding the tongue fast, both should
be pushed firmly into the tube. By blowing in
the other end of the tube the•call is produced;
the tone, degree of fineness, etc., of which is regulated by the shortened half of the plug—moving
it in or out as a finer and sharper or lower
and coarser note is required.   Some little experience and practice is, of course, required to use
it correctly.   You should always pay particular
attention to the different notes of ·wild fowl, as
well as the occasions of their being made—whether
as a call (which may be addressed either to a
mate or to a flock passing by), a note of welcome to a flock alighting, an answer to a call,
a note of suspicion, or a signal to take wing.
The call-notes especially, though also the note of
welcome, you should practise whenever you have
opportunity. When you hear them made by the

. ducks, try to imitate them as near as possible, and you will be well repaid.

Now as to why we have this good shooting here in the middle of the day : The ducks, you must know, feed mostly during the night and early morning, and would stay in their feeding-places pretty generally during the day, were it not for being driven from them by the hunters, and to such quiet, out-of-the-way places as this they come simply to rest and to avoid their persecutors.   Two or three days' banging uses up these places for a time; but in a feeding-place like the rice-pond one may shoot for weeks, of course varying his stand frequently, as the ducks would naturally avoid a point where a hunter was always to be found.   When driven from one of these roosting-places or mid-day resorts by continuous shooting, they congregate in another, which the duck-shooter should know and repair to.   That big batch of ducks that came in a short time ago was probably put out of some other slough by some trapper or duck-shooter. This I judge by their quacking and the scattered, irregular manner in which they came, as well as by their large numbers.   These smaller flocks and single ducks often get up and come back of their own accord, thinking (1 am satisfied ducks, and in fact all animal kind, have the faculty of thinking in some degree) we had gone on after putting them out.

To know how and where to find the ducks at mid-day, is one of the main requirements for noted success in duck-shooting. A thorough knowledge of the country is a great advantage; but if a person will only keep his eyes open, and take note of what is going on around him, though in a strange place, he may find game with more certainty than a blinkard who knows every foot of it.

Yes, I see them. Steady, now! I'll hold you answerable for both. Beautiful! See that puff of feathers floating down like flakes of snow! Here comes another pair. I'll try one of these. Well done again! Two more pretty ones! Quiet, Jack; he's dead enough, you greenhorn. Let's see, where was 1? I was going to tell you how to find these places. Well, then, bear this rule in mind, and make a practice and habit of it: When shooting in the morning, or in fact at any time of day, keep your eyes open, and be sure you see every duck that comes within see- ing limits. More: observe distinctly the directions of their flights, from what parts they come, and towards which points they mainly appear to be going. You will perhaps see a great many pitching down or lowering their flight over some particular place, especially late in the morning, while few or none seem to be getting up from there. Mark that carefully; that is a mid-day roosting-place. Others you see earlier in the morning, coming and going continually to and from another place. This is probably a feeding-pond. Again, you may observe a point by which the main body appears to take its flight. There is a good passway for flight-shooting. Still again, you notice a certain point where most ducks, on approaching it, appear to suddenly dart upwards and scatter quickly. You may be assured there is a shooter there. In the evening the ducks will be seen coming from the roosting-ponds and going to the feeding-grounds. Thus you may know, while shooting one day, where to look for them on the morrow. The uninitiated, non-observing numskull depends entirely on fickle luck, and probably spends the day, when not shooting, in fingering the locks of his gun, playing with or talking to his dog, or other thoughtless proceedings, and paying no attention to the flight of any ducks, unless it be the few that happen to come within range. Let me impress this on your mind: Keep your eyes open.

Mark! A pretty flock. Now see if you can't get two the first barrel. I'll wait for you. Bravo! One is crippled, but that is no discredit. A man can hardly be expected to kill more than one " clean " at a time. That is another trick whereby the skilful hunter makes the most of his opportunities; but you needn't try it too often at first. It is better for the novice to content himself with one bird at a shot. He will do better, and his shooting will improve faster. By trying to get two he often fails to get any. When you have had more experience, you can watch for them to cross, and be ready to avail yourself of those chances.

———————

It will be tedious to the reader to particularize the shooting further. Suffice it to say, after a couple of hours more of the choicest sport, we gather up our game, seventy-two head in all, making a total for the day of one hundred and twenty-five, buckle it in the game-straps, and start for the boat, which we reach without noteworthy incident. It being nearly sundown, we decide not to try evening shooting, which we would have to go to a feeding-pond again to enjoy; so we paddle directly home, where we arrive in time for a nice warm supper with "the folks."

The game-strap, an exceedingly useful part of the duck-shooter's outfit, consists of a piece of sole or heavy harness-leather, three inches broad by six inches in length, to each end of which is attached a narrow strip of bridle-leather, about two feet in length, punched at intervals of half an inch to receive the tongue of a buckle, which is fastened one at each end of the broad strip. One-half the ducks to be carried are bunched heads together, one strap buckled tightly around their necks, and the remainder being buckled to the other side, they are thrown across the shoulders, and in this way may be carried a long distance. Twenty mallards on each side make a good strapful.

CHAPTER X.

THE general character and methods of procedure in the " evening shooting " of mallards are so
similar to those described in the foregoing chapters that a full explanation of them is unneces
sary. The stand is commonly selected in some
one of their various feeding-ponds, or on some
favorable point on their route to them from
their mid-day resorts. The first named, however,
is generally the best, if it can be reached conveniently, as the shooting lasts much longer—
often until too dark to see to shoot—from the
ducks flying about the pond for some time after
coming in; and more opportunities for close shots
are to be had here than on the passways, as
their flight is lowered immediately on reaching
the pond.

After it begins to get dark there is no need
of a blind ; just stoop a little—the ducks will
take no notice of you — and by facing the west
you may see them distinctly against the light

sky long after the sun goes down, and when it would be impossible to see them in any other direction. Now you must be in readiness for snap-shots, and instantly, on seeing a duck within range, throw up your gun and pull without hesitation. If you have pointed correctly, you will be gratified by hearing the splash of your game in the water, following the report of the gun. If in a favorable locality, the numbers of the ducks and their continual quacking and whizzing by cannot fail to confuse the beginner, who will frequently stand still, undecided which one to shoot at. Experience will, however, quickly cure him (unless he be naturally of a very excitable disposition), so that no amount of game will afterwards disturb his composure.

If you have a good dog with you, and are in a place that will admit of it, let him retrieve your ducks as fast as killed; if not, and you are shooting from a boat, you must quit shooting before it gets too dark, and pick up your game, or, if you intend to come back in the morning, let them remain until then. The owls and minks, however, in such case, will very likely have robbed you of a couple or so. If you intend to shoot in the same place next morning or evening, you had better quit and pick up early, or the flashes from your gun after dark will so alarm the ducks that they will forsake it for some other less dangerous place. Many shots will frequently be had at wood-duck, teal, and sprigtails in this sport, more particularly in the rice-ponds, their common favorite feeding-places.

Now, a hint or two as to picking up your ducks in a rice-pond. Before you leave your blind fasten to the top of the tallest handy stalk of rice a piece of paper, rag, or other conspicuous object, to serve as a guide to direct your course, which you may be the better able to judge when you have gone far enough in any direction. If you have but few ducks down, it will be better to go direct to each one; but if there be fifty, sixty, or more, take a direct course from your blind, in the direction you suppose the greater part to lie, to a distance which will include all the dead in that direction, and, keeping a sharp lookout in front and on both sides, pick up all you may see. When you have reached your limit of distance, turn squarely to the right or left from fifteen to twenty-five feet, according to the density of the rice, and then take a course back towards your blind parallel to your first one.

If there are ducks beyond the blind on that side, keep on by as far as you think proper, and so continue in this way until you have been over, in parallel lines, all the circle liable to contain your ducks. Very often the ducks will drop into a mat of rice which has been broken down to the surface of the water, in which case, if they go down through it, they are easily overlooked; so be careful in your search, and if you see the tip of a wing or foot sticking up anywhere, satisfy yourself whether there is the rest of the duck fastened to it or not.

Where the rice is thick, the water not too deep, and the mud tolerably firm, you may get out of your boat, being careful to stand on the roots of the rice at first, and holding an oar near its middle with both hands, about two feet apart. Raise the oar horizontally over your head; now, bending your body only at the hips, bring the oar down forcibly against the stalks of the rice, thus breaking them to the surface of the water. Raising the oar again, step forward, with the toes turned well outwards, across and upon the prostrate stalks, and bring down the oar again as before. In this way you break your road before you, and at the same time provide a foundation which will prevent your sinking in the mud as you walk. It is rather hard work for one not accustomed to vigorous exercise, but in such a place one may travel much quicker in this way than by any other. Your ducks should be pocketed as fast as recovered, and afterwards deposited in piles at convenient places, when you may then gather them in your game-strap, and take them to the boat. The oar should be used as a cane to steady yourself when carrying the ducks on your shoulder.

CHAPTER XI.

As the cold weather approaches and the ponds begin to skim over with ice, the mallards in many places, instead of migrating further south, betake themselves to the rivers, where they congregate in large numbers, and by the combined warmth of their bodies and the constant agitation of the water manage to keep large surfaces from freezing over, long after the surrounding water has frozen to a depth of several inches. For food they depend almost wholly upon corn, which they steal from the adjacent fields, making usually two trips a day for the purpose, and extending them frequently to a distance of six or eight miles from the river. At this time they feed almost entirely by day, returning to the river to drink and to roost at night.

Though their numbers are sometimes almost incredible, comparatively few are killed in the fields, on account of the large extent of their feeding-grounds and their natural shyness when over the

land, which causes them to fly high until ready
to alight, when they circle about and lower
gradually near the centre of the field, approaching
within gunshot of the fences as rarely as possible;
and this, from the large size of the Western corn-
fields, they can easily do.

During a heavy snow-storm is the best time for
making a large bag, as the snow covers up the
corn,* which being harder for the ducks to find,
they fly lower and more continually.   At this
time the shooter's dress cannot be too white, and
he will need but a small blind (the smaller the
better in a corn-field, as the ducks will notice
any unusual appearance and avoid it).   In building it (of corn-stalks, of course) the stalks should
be stood on end, leaning against each other, and
a heavy, solid look avoided as much as possible.
After the shooter has secured a few ducks it will
be a good plan to set them up for decoys on split
corn-stalks, taking care to select as clear a space
as possible, where the stalks are low and thinly
dispersed.   As they get covered with snow they
should be shaken clean again, or they will be of

* The ducks usually feed on the scattered grains lost in husking, or
the small ears thrown away, rather than tear the husks off themselves.

no advantage. The usual call-note is never to
be practised here, as it will only serve to frighten
the ducks, who never call when feeding in the
fields, but make a kind of low, chattering noise,
which from its sound seemingly implies content-
ment and happiness.

Large shot, No. 2 or 3, Leroy's size, with heavy
charges of powder, should be used, and aim taken
at a thick bunch, if possible, where the chances
for breaking a wing or striking the head and neck
are increased. Wing-broken ones are usually easy
to secure, being killed by the fall, if the snow is
not deep; and if not killed, they may be easily
tracked upon it, if running away. One of the chief
objections to this sport is its cruelty, so very many
ducks are hit that fly on to linger perhaps for
days in agony until relieved by death.

Long shots are occasionally to be had at geese
and brant in the fields, as they frequently remain
roosting and feeding with the mallards long after
winter sets in. The brant are generally the first
to leave, the geese next, the mallards staying to
the last, some of them frequently the entire winter, after the holes in the river freeze over,
roosting in the small spring-holes and creeks fed
by the same.

The sportsman will often happen on a bevy of quail or pack of prairie-chickens in his corn-field excursions, and thus, if he be accompanied by a good dog, may enjoy a variety of sport. He should be very careful not to leave his game-strap at home.

CHAPTER XII.

"Big counts" are frequently made at the holes in the ice where the ducks roost and come to drink, and also at the shallow, open water at the mouths of spring creeks whose bottoms are covered with sand or gravel, and which the ducks seek as aid in digesting their food. This sport seldom lasts very long, as the air-holes freeze over quickly in cold weather if the ducks are kept out of them, and the ducks are thereby forced to leave and hunt open water elsewhere. But few directions, not heretofore given, are necessary to the novice in this branch of the sport. The providing of the blinds has been described in the chapter relating especially to them. All the decoys that can be conveniently procured should be used, the dead ducks being set on the ice, near the edge of the hole, with their heads upon stakes or under their wings. There is considerable danger attending the sport, however, and it should never be attempted until the ice is sufficiently strong to prevent the possibility of breaking through.

As it is sometimes necessary for the wildfowler to cross weak ice, I will give a few precautions which it will be well for the novice to observe. Keep as close to your boat as possible at all times; and instead of walking in front of the boat, and dragging it after you by the chain, lay hold of it with the hands on each side, about two and a half or three feet from the stern, and thus push it before you; then if you break through the ice, you will fall upon the stern of the boat instead of into the water. A light pole six or eight feet in length, with a sharp iron point in the end, is very useful on weak ice. The fowler may then remain in his boat, and propel it by pushing the sharp end of the pole against the surface of the ice. If the boat should break through he should stand near the stern so as to lift the bottom of the boat at the bow above the surface of the ice, and either push against the edges of the firm ice with the pole, or use a paddle in the open water at the side of the boat. The boat should then be " rocked " continually to break the ice as it goes. And when new ice is found that the boat will not break, the fowler, after pushing the boat as far out of the water as possible, should step quickly to the bow, and, resting one knee upon it, should push with the other leg against the ice until he can get out, and push again from the stern, or use his pole as first described.

If you have no boat, and must cross where you suspect the ice to be weak, cut a strong pole a couple of inches thick and eight or ten feet long, and then keeping hold of the pole with both hands to prevent your sinking entirely in case the ice should break, lie down at full length on your breast, and you may wriggle across safely where otherwise it would be impossible. If you should break through, and have no companion with you, don't be frightened, don't tire yourself out with useless struggles, breaking ice in all directions, but take time to determine your best course and shortest way out, and then stick to it. By so doing you will easily get out where a nervous, excited person would inevitably be drowned.

At the mouths of the creeks and shoals where the ducks come for sand, large numbers are frequently killed from a blind built at a convenient

distance on the shore. I had, together with my
partner, most excellent sport for several days one
winter at two such places about a mile apart,
both shooting on alternate days at either place.
The ducks at this time were feeding on corn,
and would come in from the fields with their
craws completely distended—often as much as a
gill in each one. Those of the dead ones, after
we had finished shooting, we would cut open and
scatter the corn from them about in the shallow
water, and the ducks finding it there next day,
while we were shooting at the other place, would
be so tempted by the bait as to almost make it
hard work to keep them away when the third
day came. Decoys should be used, " the more
the merrier "; and " call your prettiest " whenever
you see a duck passing.

TIMBER MALLARD SHOOTING—SPRING.

CHAPTER XIII.

MALLARD SHOOTING IN THE TIMBER—SPRING.

In the spring, when the heavily-timbered "bottom-lands," as they are called, are inundated by the rising of the rivers, mallards may be found sitting in large bodies, both by night and day, in the depths of the woods, particularly amongst the maple and willow timber, where they feed on the larvæ, buds, and vegetable matter found there floating on the surface of the water. On being routed, instead of settling in some other place, and there remaining for the day, they will come back shortly, singly and in small parties, affording most excellent sport. This fact is unknown to many sportsmen, who think, as I did in my earlier duck-shooting days, that the ducks cared but little more for one place than another, and even if they wished to come back could not, or at least would be very unlikely to, find the place again after once leaving.

If the water is shallow enough to admit of

wading, unless there happens to be a fallen tree-top or pile of brush in a favorable position for shooting, and large enough to conceal both boat and shooter, the sportsman would better, after find-ing a place for his boat where it will not be seen, get out and take a position as near as pos-sible to where the main body of the ducks are sitting. He must avoid standing under large branches or an overhanging tree-top, as such would interfere with his shooting; but if he can find an old log, stump, or clump of bushes in the proper location, he should get behind it, though if his dress be of the proper color, and he will hide his face and avoid moving suddenly when ducks are approaching, a blind is not absolutely neces-sary. On seeing ducks flying about, no matter in what direction, he should call loudly; and after-ward from time to time repeat the call, whether ducks are in sight or not. Frequently they may be in hearing when he is unable to see them through the thick woods; and in no place will ducks answer the call as readily as here. De-coys are seldom used, yet if there is an open place handy where they may be readily seen, it is a good plan to stick up a few dead ones. A retriever is necessary for this sport, particularly

if the water is deep—up to the calf of the leg
or so—for if the sportsman is compelled to secure his cripples alone, he may catch his toe
under a hidden root or stick when in pursuit of
them, and thereby perhaps experience a style of
"ducking" not looked for and little to be desired. Sometimes, too, he may find good shooting, as I have done, when the buck-brush is so
close that the boat cannot be easily pushed
through it, and the water perhaps just too deep
to wade. In such a case, if he has no retriever,
he must leave it and look up another place.

It is better to collect the dead ducks as fast
as killed, as by so doing he will be able to
shoot until the last possible moment, not having
to quit before dark to find his game.

Mallard are said by some authors to dive occasionally for food in the spring. Though they
may, I have never seen them do so, and think it
rather unlikely.

Timber mallard shooting is one of the best
of sports. The ducks come sailing so slowly
about among the trees, with wings extended,
that the veriest novice can hardly fail to hit
them, and the experienced sportsman will usually
kill his "right and left" easily. To give the

reader an idea of what may be done in this sport, I subjoin a memorandum of shooting done by a friend of the author, Mr. F. Kimble, a genuine duck-shooter, during the spring of 1872, all with a single-barrelled muzzle-loading gun, 9 gauge. Not over three ducks were killed at any one shot, and nearly all singly :

Feb. 27, killed 70 ducks.	Mar. 9, killed 82 ducks.
" 28, " 74 "	" 10, " 60 "
" 29, " 81 "	" 11, " 72 "
Mar. 1, " 76 "	" 12, " 128 "
" 2, " 106 "	" 13, didn't shoot.
" 3, " 61 "	" 14, killed 122 ducks.
" 4, didn't shoot.	" 15, " 70 "
" 5, killed 66 ducks.	" 16, " 68 "
" 6, " 107 "	
" 7, " 57 "	Total, 1,365 "
" 8, " 65 "	

Total 17 days' shooting, 1,365 ducks, and 5 brant not included in memorandum. His ammunition gave out almost every day. Not expecting to find such a large amount of game, the party he was with took but little with them, and the "store-keeper" at the little town near by would order only a keg or so of powder at a time, and then would not sell it all to one person at any price, for fear of offending others.

Besides these haunts already mentioned, mallard are very partial to overflowed prairies and grainfields, the shallows among low willows and pin-oaks, whose tiny acorns they are particularly fond of. But it is unnecessary to characterize the shooting of them further; the various suggestions already given are applicable to their successful pursuit in all places.

CHAPTER XIV.

BLUE-WINGED TEAL (ANAS DISCORS).

ADULT MALE: bill almost as long as the head, deeper than broad at the base, depressed towards the end, its breadth nearly equal in its whole length, being, however, a little enlarged towards the rounded tip.

Head of moderate size, oblong, compressed. Neck of moderate length, rather slender. Body full, depressed. Feet short, placed rather far back.

Plumage dense, soft, and blended. Feathers of the head and neck very small and slender; of the back and lower parts in general broad and rounded. Wings of moderate length, rather narrow and acute. Primaries strong, slightly curved, tapering; the first scarcely longer than the second; the rest rapidly decreasing. Secondaries broad, the outer obliquely rounded, the inner elongated and acuminate, as are the scapulars. Tail short, rounded, and acuminate, of fourteen rather narrow acuminate feathers.

182

Bill bluish-black. Iris dark hazel. Feet dull yellow; webs dusky; claws brownish-black, with the tips grayish-yellow. Upper part of the head black; a semilunar patch of pure white on the side of the head before the eye, margined before and behind with black; the rest of the head and the anterior parts of the neck of a deep purplish-blue with purplish-red reflections; the lower hind neck and fore part of the back brownish-black glossed with green. Each feather with a curved band of pale reddish-buff, and a line or band of the same in the centre; the hind part of the back greenish-brown, the feathers edged with paler. The smaller wing-coverts of a rich ultramarine blue, silky, with almost metallic lustre. Alula, primary coverts, and primary quills grayish-brown, edged with pale bluish; outer secondaries of the same color, those of the speculum duck-green, changing to blue and bronze, with a narrow line of white along their terminal margin; the inner greenish-black on the outer web, greenish-brown on the inner, with a central line and narrow external margin of pale reddish-buff, the more elongated scapulars similar, but some of them margined with greenish-blue. Secondary coverts brown, with their terminal white. Tail feathers chocolate-brown, slightly glossed with green, their margins buffy. The lower parts are pale reddish-orange, shaded on the breast with purplish-red, and thickly spotted with black; the number of roundish or elliptical spots on each feather varying from ten to twenty-five, those on the upper and hind parts of the sides running into transverse bars. Axillar feathers, some of the lower wing-coverts, and a patch on the side of the rump pure white; lower tail-coverts brownish-black.

Length to end of tail, 16 inches; extent of wings, 31¼; weight, 12½ ounces.

ADULT FEMALE: Bill greenish-dusky. Iris hazel. Feet of a duller yellow than those of the male. The head and neck are pale dull-buff, longitudinally marked with brownish-black lines, which are broader and darker on the top of the head. The fore parts of the cheeks and the throat whitish, without markings. The upper parts are dark-brown. The feathers margined with brownish-white. The smaller wing-coverts colored as in the male, but less brilliantly. No blue on the scapulars, which are also less elongated. On the lower parts the feathers are dusky-brown, broadly margined with light brownish-gray, of which there is a streak or spot in the centre. The axillary feathers and some of the lower wing-coverts are white; but the patch of that color so conspicuous in the male is wanting.

Length to end of tail, 15 inches; extent of wings, 24; weight, 10½ ounces.

The flesh of the blue-winged teal is considered by epicures to be superior in delicacy of flavor to that of most water-fowl, that of the red-head and canvas-back alone excepted; and as they are seldom found in poor condition, unless having been previously wounded, generally command a high price in market, and are consequently much sought for by the market-hunters. They are one of the very first of the duck tribe to make their appearance from the north, and congregate in vast numbers on their favorite feeding-grounds of the Western country. The seeds of the wild oats or rice, and grass, various kinds of pond-weeds, and mosses, and insects, are accepted by them as food. They are very partial to small, muddy-bottomed streams whose shallow edges are bordered with wild rice, and the broad leaves of the pond-lily, under which they are often to be seen sitting, seemingly to escape the too intense glare and heat of the sun; and are always to be found too, in proper season, in the shallow ponds or sloughs whose slimy, stagnant waters are filled with the dense growth of weeds and mosses; but are seldom or never met with in ponds or streams having gravelly or sandy bottoms. The common method of hunting them is similar to that described under the heads of morning and evening mallard-shooting, but great numbers are killed by stealing upon them silently when feeding. A very slight blow brings them down, and, as they usually sit quite close together, as many as fifteen or twenty are often killed at a single discharge of a common shoulder gun. Being tamer and less wary than most other water-fowl, they may be easily approached with ordinary caution. Small shot should be used for shooting them, No. 7 or 8 being the proper size. They may be readily called by imitating their notes properly, which are very similar in character and expression to those of the mallard, but shorter and pitched higher, such as a mallárd of the same size and proportionately feeble constitution might be supposed to make. They never appear to be suspicious of decoys, but drop amongst them immediately without hesitation. And as they are always found in shallow water, the dead ones may be set up for decoys, as described on

page 68 in mallard shooting. Bear in mind to
·set the decoys in such a position that they may
show to best advantage ; advice on the subject
of decoying will be found in the chapter under
that head.

Capital sport may be had " jumping them,"
as it is called, after they are routed from their
feeding-ponds and during the middle of the day,
when they may be found sitting in small parties
or singly in the edge of the rice which borders
some favorite stream. A very light paddle-boat
is generally used. If two go together, one shoots
while the other propels the boat; it is usual,
however, for one to hunt by himself, in which
case he must, of course, do his own paddling.
He should sit or kneel near the stern, with his
gun in front of him, and in proper position to
be seized instantly on the bird rising, and paddle
quietly yet quickly, keeping as near the edge of
the rice as possible, yet taking care not to allow
the boat to graze any of the stalks, which might
alarm the game. He should be constantly on his
guard to detect any movement of his game, and
instantly, on the bird's rising, he should drop
the paddle, no matter whether in or out of the
boat, and be ready to shoot as soon as need be.

The paddle should be fastened to the boat by a short, light cord, so that it may be regained· easily if dropped overboard. When approaching teal on the water with the intention of shooting, fire as soon as sufficiently near, as they never give warning when intending to rise, but spring at once, irrespective of position. They seldom fly far on on being flushed, unless continually persecuted, and alight very much in the manner of wood-cock. Their flight is usually low and very fast; and when in flocks, packed closely together, they afford fine opportunities for killing several at a shot. The tyro will be frequently surprised at the small number killed from a large flock, if fired at at the wrong moment. They should be allowed to pass a little, and never fired at as they are approaching, for in such case at the report of the first barrel they instantly scatter in all directions, making the second barrel of but little use. They come together again, however, in a few rods' flight. They are not very expert divers, but will endeavor to secrete themselves, if wounded, and will remain perfectly quiet, often allowing the hunter to almost step upon them before moving. In common with all ducks, they have the power of sinking the body level with the surface of the water when wounded, and often escape in this manner if the water is at all rough. In spring they are rarely seen in large numbers.

CHAPTER XV.

GREEN-WINGED TEAL (ANAS CRECCA)..

ADULT MALE : Bill almost as long as the head, deeper than broad at the base, depressed toward the end, its breadth nearly equal in its whole length, being, however, a little enlarged toward the rounded tip.

Head of moderate size, compressed. Neck of moderate length, rather slender. Body full, depressed. Wings rather small. Feet short, placed rather far back. Claws small, curved, compressed, acute; the hind one smaller and more curved; that of the third toe largest, and with an inner sharp edge.

Plumage dense, soft, blended. Feathers of the middle of the head and upper part of hind neck very narrow, elongated, with soft, filamentous, disunited bands; of the rest of the head and upper parts of neck, very short; of the back and lower parts in general, broad and rounded. Wings

190

of moderate length, narrow, acute. Tail short, rounded, and acuminate, of sixteen acuminate feathers.

Bill black. Iris brown. Feet light bluish gray. Head and upper part of the neck chestnut-brown; a broad band narrowing backward from the eye down the back of the neck, deep, shining green, edged with black below, under which is a white line, which, before the eye, meets another that curves forward and downward to the angles of the mouth. Chin brownish-black, as are the feathers at the base of the upper mandible. Upper parts and flanks beautifully undulated with narrow brownish-black and white bars; anterior to the wings is a short, broad, transverse band of white. Wings brownish-gray; the speculum in the lower half violet-black, the upper bright-green, changing to purple, and edged with black; behind margined with white, before with reddish-white. Tail brownish-gray, the feathers margined with paler; the upper coverts brownish-black, edged with light yellowish-gray. Lower part of neck anteriorly barred as behind. Breast yellowish-white, spotted with black; its lower part white. Abdomen white, faintly barred with gray. A patch of black under the tail; the lateral tail-co-

verts cream-colored, the larger black, with broad
white margins and tips.

Length to end of tail, 14¾ inches; extent of
wings, 24; weight, 10 ounces.

ADULT FEMALE: The female wants the elongated
crest, and differs greatly in coloring. The head
and neck are streaked with dark-brown and light-
red; the fore-neck whitish; the upper parts mottled with dark-brown; the anterior feathers barred; the posteriors margined with yellowish-
white. The wings are nearly as in the male,
but the green of the speculum is less extensive.
The lower part of the fore-neck is tinged with
yellowish-red and mottled with dark-brown, as
are the sides; the rest of the lower parts white.

Length to end of tail, 13¾ inches; extent of
wings, 22½; weight, 10 ounces.

This variety resembles the preceding very generally in form, habits, and manner of flight, and
its flesh is considered of nearly equal delicacy.
They are a trifle smaller than the blue-wings,
and their plumage is more varied and beautiful.
Though generally making their appearance in the
fall at about the same time, they are of a more
hardy, enduring disposition than the other variety, and remain much later, often until the wea-

ther gets very cold. In flight and upon the feeding-grounds the two associate together promiscuously. Unlike the blue-wings, the green-wings are quite tenacious of life, and are more expert in diving. Their call-notes, too, are entirely different, being a succession of short, sharp whistles (pitched about " high F " on a piano), by imitation of which they may frequently be decoyed within gun-shot, as they may also be by means of stools, or wooden decoys.

In the spring, though often quite abundant, they are seldom made the special objects of pursuit by the hunters, the larger and more profitable species of wild fowl then claiming their attention. If the sportsman is desirous of shooting them, however, he should visit the low, overflowed bottom-lands, where, amongst the low willows and buck-brush, they are almost certain to be found, the tender buds of the willow and other trees being their principal food at this season. They frequently resort, too, to overflowed grassy prairies, and feed upon the seeds of the grass which float upon the surface of the water. In no other branch of wild-fowling is a breech-loader of more advantage than in teal-shooting, in consequence of the large number of cripples often to be secured after a successful shot. Rapidity of loading is then of especial importance.

Finally, they are generally considered most pleasing to the epicure when properly broiled and served in becoming style.

CHAPTER XVI.

Local names : " Sprigtails," " Sharptails," and " Water-Pheasants."

ADULT MALE: Bill nearly as long as the head, deeper than broad at the base, depressed toward the end, the frontal angles short and obtuse.

Head of moderate size, compressed, the forehead rounded. Neck rather long and slender. Body full and depressed. Wings rather small. Feet very short, placed rather far back; tarsus very short, compressed at its lower part.

Plumage dense, soft, blended. Feathers of the head and neck short; on the hind head and neck elongated. Wings narrow, of moderate length, acute. Tail of moderate length, tapering, of fourteen tapering feathers, of which the two middle project far beyond the rest.

Bill black; the sides of the upper mandible light-blue. Iris brown. Feet grayish-blue. Claws black. Head, throat, and upper part of the neck anteriorly greenish-brown, faintly margined behind

195.

with purplish-red. A small part of hind neck dark-green; the rest, and the upper parts in general, beautifully undulated with very narrow bars of brownish-black and yellowish-white. Smaller wing-coverts, alula, and primary quills gray, the latter dark-brown towards the end. Speculum of a coppery-red, changing to dull green; edged anteriorly with light brownish-red, posteriorly with white. The inner secondaries and the scapulars black and green, with broad gray margins. Upper tail-coverts cream-colored, the outer ribs blackish and green; tail light-gray, the middle feathers dark-brown, glossed with green. On each side of the neck is an oblique band of white, of which color are the upper parts in general; the sides, however, are undulated like the back; the lateral feathers of the rump cream-colored; the lower tail-coverts black, those at the sides edged with white.

Length to end of tail, 29 inches; extent of wings, 36; weight, 2 pounds.

Adult female: The female, which is much smaller, has the upper parts variegated with brownish-black and light yellowish-brown; the margin of the feathers and a mark on each side of the shaft being of the latter color. The speculum is dusky green, margined behind with white. The primary quills grayish-brown. The lower parts are of a light brownish-yellow, the sides variegated with brown; the bill is black; the iris brown; the feet light bluish-gray.

Length, 22½ inches; extent of wings, 34; weight, 1 pound 9 ounces.

The sprigtail, the most graceful and symmetrically formed of the whole duck tribe, like the mallard, is found in nearly every State in our Union, with the exception of Maine and the New England States in general, and it is often made a cause for wonder amongst sportsmen that they do not frequent these States more, and Maine more especially, as it would seem its hundreds of lakes and streams, with acres of marsh-lands adjoining, might prove enticement sufficient; but simply because their proper food is not to be found in these lakes or thereabouts in needed abundance is the sole and ample reason for their non-appearance. The timber of Maine consists chiefly of pine; oak is very scarce, and pin-oaks, whose tiny acorns are greedily sought for by mallards and sprigtails, are unknown. Smart-weed, cockle-burrs, and wild oats never grow there, and corn is not raised in sufficient plenty to afford food for them.

Beech-mast, a favorite food of sprigtails, is often very abundant, but the trees are as often on the hills far from water as near by, and no duck often makes a business of looking for food many miles from water, unless it be to sometimes steal a little corn or other grain from the farmers. The food of the black ducks, which are often found there in quite goodly numbers, consists mainly of leeches, snails, insects, and larvæ; and though sprigtails and mallards often partake of them, I am inclined to believe they would prefer wild oats as a steady diet.

Sprigtails are not usually so plenty in the Western States in the fall as during the spring; but a few make their appearance during September, associating very generally with the other shoal-water ducks, but with the mallard most particularly; feeding and travelling with the same flock continually for days, and timing their flights, which are usually much faster than that of the mallard, to its rate of speed. Upon the breaking up of winter, however, they begin to arrive in countless numbers, taking possession, as it were, of the overflowed prairies and corn-fields, where they feed upon the previous season's waste and unharvested grain, and the grass-seeds which, floating upon the surface of the water, become drifted together into large patches. Here they soon become exceedingly fat and their flesh fine-flavored. They fly closely together without order, darting aimlessly about, and it is not uncommon to kill several at a shot. When travelling, their flight is steadier, and they often keep up a continual cackling and whistling noise. Their call-note is a low, plaintive whistle of one tone two or three times repeated, which they will answer readily upon its being well imitated; but it is as well to use the mallard call for this fowl, as they answer it full as well, and decoy to mallard-stools as quickly and readily as though of their own kind. It is therefore unnecessary to make decoys to imitate sprigtails. They do not show as well as mallard decoys, being smaller and the colors more neutral and unattractive; and if made with long necks, as the natural birds are, they are easily broken, besides taking up too much room when moving about.

In sprigtail-shooting it is best to place the decoys to windward of the blind when circumstances will allow, particularly when the ducks are shy; and in this respect they differ from all other wild-fowl. When much pursued, they get

very wary, and often when shooting mallards I
have had old cock-sprigtails come up against the
wind, almost within gun-shot, when they would
"jump" back all at once and circle about to leeward of the decoys, and, after coming and going
perhaps half a dozen times, finally drop down just
out of reach. I never saw a hunter who did not experience a most happy feeling of satisfaction when
he had succeeded in killing such a one. The object
of putting the decoys to windward is to take advantage of this habit of circling about to their leeward. When in flocks, they generally decoy much
better and appear less suspicious.

When wounded, they endeavor to escape by
running, if on land, and will frequently hide and
crouch motionless to escape observation. They
are not very expert divers, and when wounded
usually try to remain under water so long as to
quickly tire themselves out, when they may be
easily captured. In the fall they remain until the
ponds are skimmed over with ice, when they
take up their departure for the south. No. 5
or 6 St. Louis shot is the best size for ordinary
sprigtail-shooting.

CHAPTER XVII.

THE WOOD-DUCK OR SUMMER DUCK (ANAS SPONSA).

ADULT MALE: Bill shorter than the head, deeper than broad at the base, depressed toward the end, slightly narrowed toward the middle of the unguis, the frontal angles prolonged and pointed.

Head of moderate size. Neck rather long and slender. Body full and depressed. Wings rather small. Feet very short, strong, placed rather far back, tarsus very short, considerably depressed.

Plumage dense, soft, blended, generally glossed. Feathers of the middle of the head and upper part of the hind neck very narrow, elongated, and uncurved; of the rest of the head and upper part of the neck very short; of the back and lower parts in general broad and rounded, excepting on the shoulders before the wings, where they are enlarged, very broad, and abrupt. Wings of moderate length, narrow, and acute. Tail of moderate length, rather broad, much rounded, of sixteen rounded feathers.

Upper mandible bright-red at the base, yellowish at the sides; the intermediate space along the ridge and the unguis black, as in the lower mandible and its membrane. Iris and edges of eyelids bright red. Feet dull orange. Claws black. Upper part of the head and space between the bill and the eye deep green and highly glossed; below the latter space a patch of dark purple, and a larger one of the same color, but lighter, behind the eye; sides of the neck, its hind parts under the crest, and the middle all round very dark purple. Throat for more than three inches pure white, with a process on each side a little beyond the eye, and another nearly half way down the throat. Sides of the neck and its lower part anteriorly reddish-purple; each feather on the latter with a triangular white tip. Middle of the neck behind, back, and rump very dark reddish-brown; the latter deeper and tinged with green. Upper tail-coverts and tail greenish-black. Some of the lateral tail-coverts dull reddish-purple; a few on either side with their filaments light-red. Smaller wing-coverts, alula, and pri-

maries dull grayish-brown.   Most of the latter,
with part of their outer web grayish-white, and
their inner, toward the tip, darker and glossed
with green. Secondary quills tipped with white,
the outer webs green, with purple reflections.
Those of the inner secondaries and scapulars
velvet-black, their inner webs glossed and changing
to green.   The broad feathers anterior to the
wings are white, terminated with black.   Breast
and abdomen grayish-white.   Feathers under the
wings yellowish-gray, minutely undulated with
black and white bars.   Lower wing-coverts and
axillar feathers white, barred with grayish-brown.
Lower tail-coverts dull grayish-brown.

Length to end of tail, 20½ inches; extent of
wings, 28.

ADULT FEMALE : The female is considerably
smaller, and differs greatly from the male in
coloring. The feathers of the head are not elongated, but those of the upper part of the neck
are slightly so. In other respects the plumage
presents nothing very remarkable, and is similar
to that of the male. Bill blackish-brown. Feet
dusky, tinged with yellow.   Upper part of the
head dusky, glossed with green.   Sides of the
head and neck, and the hind part of the latter,

light brownish-gray.   Throat white, but without
the lateral processes of the male.   Fore part of
the neck below, and sides, light yellowish-brown,
mottled with dark grayish-brown, as are the sides
under the wings.   Breast and abdomen white,
the former spotted with brown.   Hind neck, back,
and rump dark-brown, glossed with green and
purple.   Wings as in the male, but the speculum less, and the secondaries externally faint
reddish-purple ; the velvet-black of the male di-
minished to a few narrow markings.   Tail dark-
brown, glossed with green.

Length, 19½ inches.

The wood-duck is the most beautifully plu-
maged of the whole family of wild fowl.   They
are common to nearly all parts of our Union,
excepting the sea-coast, which they rarely visit.
"They build their nests frequently in hollow
trees and stumps, and from this circumstance
probably received their name."   They never dive
for food, but live chiefly upon acorns, pond-
moss, the seeds of the wild-oats, and insects,
and are to be found, too, feeding on the scattered and waste kernels of wheat and other grain
which are always left upon the fields after harvest.

The most successful method of hunting wood-ducks, and that most in use amongst hunters, is identical with that described under the head of blue-winged teal, *i.e.*, "jumping them up" along the creeks and rivers which they frequent. The best season for this sport is the latter part of August and the month of September. In this sport watch carefully about the old logs and rat-houses which are to be found along the edges of the reeds. Wood-ducks are very fond of sitting on such places during the middle of the day to preen themselves and bask in the sun.

Their usual note (this varies considerably, however) is a sound between a whistle and a squeal, commenced in a medium tone, and finished about three notes higher, slurring gradually, though not smoothly, the intermediate tones. They are not easily decoyed, either by stools or calls. In the fall, just about dusk in the evening, they frequently gather in quite large flocks in some sheltered bend of the river or in some favorite pond to roost. They fly through the woods very rapidly, darting about amongst the trees much like wild pigeons. Are not very tenacious of life, but when crippled are very cunning. Dive very well, and if near the land usually try to escape by hiding.

CHAPTER XVIII.

Local name, "Bald-pate."

ADULT MALE: Bill nearly as long as the head, deeper than broad at the base, depressed towards the end, the sides nearly parallel, the tip rounded.

Head of moderate size, oblong, compressed. Neck rather long, slender. Body elongated and slightly depressed. Feet very short; tibia bare for about a quarter of an inch; tarsus very short, compressed.

Plumage dense, soft, blended. Feathers of the head and upper neck oblong, small; those along the crown and occiput longer; of the lower parts ovate, glossy, with the extremities of the filaments stiffish. Wings rather long, little curved, narrow, pointed. Tail short, rounded, and pointed, of sixteen feathers, of which the middle pair are more pointed, and project considerably.

Bill light grayish-blue, with the extremity, in-

206

cluding the unguis and a portion of the margins, black. Iris hazel. Feet light bluish-gray, the webs darker, the claws dusky. The upper part of the head is white, more or less mottled with dusky on its sides; the loral space and cheeks reddish-white, dotted with greenish-black; a broad band from the eye to behind the occiput deep-green. The lower part of the hind neck, the scapulars, and the fore part of the back are minutely transversely undulated with brownish-black and light brownish-red; the hind part similarly undulated with blackish-brown and grayish-white. The smaller wing-coverts are brownish-gray; the primary quills and coverts dark grayish-brown; the secondary coverts white, tipped with black. The speculum is duck-green anteriorly, bounded by the black tips of the secondary coverts. The tail feathers are light brownish-gray. The throat is brownish-black; the lower part of the neck in front and the fore part of the breast light brownish-red; the breast, belly, and sides of the rump white; the sides of the body finely undulated with white and dusky; the rump beneath and the lower tail-coverts black.

Length to end of tail, 20½ inches; extent of wings, 34½; weight, 1 pound 14 ounces.

ADULT FEMALE : The female is considerably
smaller. The bill, feet, and iris are colored as in
the male. The head and upper part of the neck
all round are white or reddish-white, longitudinally streaked with brownish-black, the top of the
head transversely barred; the lower part of the
neck in front and behind, the fore part of the
back, and the scapulars, are blackish-brown; the
feathers broadly margined with brownish-red and
barred with the same; the bars on the back
narrow; the hind part of the back dusky; the
upper tail-coverts barred with white. The wings
are grayish-brown; the secondary coverts tipped
with white; the secondary quills are brownish-
black, the inner grayish-brown, all margined with
white. All the lower parts are white, excepting
the feathers of the sides and under the tail, which
are broadly barred with dusky and light reddish-
brown.

Length to end of tail, 18 inches; extent of
wings, 30; weight, 1 pound. 5 ounces.

I have found widgeon most abundant on the
mossy, stagnant waters of Southern Missouri and
Tennessee, but never in such numbers, however,
as to warrant particular notice. There they associate indiscriminately with the gray duck or gad-

wall, the most numerous duck in those localities. They feed more by night than day, chiefly upon pond-moss, the blades, roots, and seeds of various water-grasses, insects, etc., and occasionally, though rarely, dive in shoal water to secure them. On the Chesapeake they are said to be particularly fond of the roots of the wild celery, but being poor divers, depend upon stealing their supplies from the canvas-back—a trick they are said to be very expert in. " Watching for the moment of the canvas-back's rising, and before he has his eyes well opened, they rush forward, snatch the delicious morsel from his mouth and make off to enjoy it." 1 have never seen them in company with the canvas-back in the West, but have often found them associated with various shoal-water ducks. Their " call " is a soft, plaintive whistle of two tones and three notes of nearly equal duration, the first and second of the same pitch, the third about three tones lower; the second or middle is strongly accented. They stool well to almost any decoys, to mallard perhaps better than any others, and may be decoyed by imitating their notes or those of the mallard. Their flesh is excellent eating, but they soon spoil in warm weather if the entrails are not drawn. They may be distinguished from others of the duck tribe by their proportionally greater length of wing. A slight blow brings them down, and as they fly clustered together, several are often killed at a discharge. Methods of hunting them similar to that of mallards, as before described.

CHAPTER XIX.

Local name, " Gray Duck."

ADULT MALE: Bill nearly as long as the head, deeper than broad at the base, depressed towards the end, the sides parallel, the tip rounded.

Head of moderate size, oblong, compressed. Neck rather long and slender. Body elongated, slightly depressed. Feet very short; tibia bare for about a quarter of an inch; tarsus very short, compressed.

Plumage dense, soft, blended. Feathers of the head short, of the occiput and nape a little elongated, of the lower parts glossy, with the extremities of the filaments stiffish. Wings rather long, little curved, pointed. Tail short, rounded, of sixteen strong, pointed feathers, of which the middle pair project considerably.

Bill bluish-black. Iris reddish-hazel. Feet dull orange-yellow. Claws brownish-black. Webs dusky. Head light yellowish-red; the upper part and nape much darker and barred with dusky;

211

the rest dotted with the same. The lower part
of the neck, the sides of the body, the fore part
of the back, and the outer scapulars, undulated
with dusky and yellowish-white; the bands much
larger and semi-circular on the fore part of the
neck and breast; the latter white. The abdomen faintly and minutely undulated with brownish-gray. The elongated scapulars brownish-gray,
broadly margined with brownish-red. The hind
part of the back brownish-black. The rump all
round and the upper and lower tail-coverts bluish-black. The anterior smaller wing-coverts are
light-gray, undulated with dusky; the middle
coverts of a deep, rich chestnut-red; primary
coverts brownish-gray; outer secondary coverts
darker and tinged with chestnut; the rest black,
excepting the inner, which are gray. Tail brownish-gray, the feathers margined with paler.

Length to end of tail, 21¾ inches; extent of
wings, 35; weight, 1 pound 10 ounces.

ADULT FEMALE: The female is considerably
smaller. Bill dusky along the ridge; dull yel-
lowish-orange on the sides. Iris hazel. Feet of
a fainter tint than in the male. Upper part of
head brownish-black, the feathers edged with light
reddish-brown. A streak over the eye, the cheeks,

the upper part of the neck all round, light yellowish-red tinged with gray, and marked with small, longitudinal, dusky streaks, which are fainter on the throat, that part being grayish-white. The rest of the neck, the sides, all the upper parts, and the lower rump feathers, brownish-black, broadly margined with yellowish-red. Wing-coverts brownish-gray, edged with paler; the wing otherwise as in 'the male, but the speculum fainter. Tail feathers and their coverts dusky, laterally obliquely indented with pale brownish-red and margined with reddish-white.

Length to end of tail, $19\frac{1}{4}$ inches ; extent of wings, 31.

The habits of this bird seem to have been rather imperfectly understood by several of our best authorities on sporting and natural history ; Forrester asserting it to be " a solitary bird, rarely congregating in large bodies," while Wilson describes it as being " a very quick diver, so much so as to make it difficult to be shot." Both these authors state their knowledge of this species to be very limited, so I consider their remarks to be the result of information received from others less careful in their observations or not sufficiently familiar with its ways. 1 can readily conceive how such a judgment might be formed, as the habits of wild fowl often vary in different localities. In the more northern States I have seldom seen gray ducks assembled more than four or five together, but consider this as being due to their general scarcity; for in Tennessee and Southern Missouri, the former State more especially, I have found them associated in flocks of thousands upon their feeding-grounds, more numerous than all other ducks, and, like mallards, separated into smaller flocks of various numbers when flying from one feeding-ground to another. As to their habits of diving, though having hunted them one season for three. months steadily in preference to all others, on account of their greater numbers, I have never yet seen one dive either for the purpose of feeding or to escape being shot, unless having been previously wounded, when they become exceedingly cunning, and are then as expert as the widgeon or mallard in diving. It is a favorite trick with them to seize the roots of the weeds when wounded and under water, and cling to them, if possible, until the hunter has passed on. They decoy exceedingly well to mallard decoys, and come readily to the mallard call, which resembles their own very closely.

The various methods of hunting them are very similar to those already described for other shoal-water ducks. In Tennessee, next to decoy-shooting, my favorite way was to suddenly yet cautiously come upon them in a boat from behind the numerous little points along the lake-shore, when they would be compelled to fly by me in escaping, not being able to rise so nearly perpendicularly above the high cypress timber which bordered the little coves (or pockets, as they are called by the natives) where they were almost always to be found feeding. Many opportunities, too, for sitting-shots at flocks were had in this sport, and quite large numbers were often bagged during the day. Light "dug-outs" are used by the natives of this locality for shooting from, but it is impossible to shoot broadside from them with a heavy gun without upsetting; consequently many opportunities for flying shots are lost, and for this reason they are not as useful as the regular hunting-skiff.

Gadwalls command a good price in market, and are ranked with the mallard and redhead; these, with the canvas-backs, are denominated "large ducks" by the hunters and market-men, all others being classified as small ones. Their notes and manner of flight resemble that of mallards; their food, habits of feeding, etc. (excepting diving), similar to those of the widgeon. They die easily on being struck, and are commonly tame and easy to approach. St. Louis "fives" are the proper sized shot.

CHAPTER XX.

Local names : "Spoonbills," " Butler Ducks " (the last a name of quite recent origin).

ADULT MALE : Bill longer than the head, higher than broad at the base, depressed and much widened towards the end, where its breadth is doubled.

Head of moderate size, oblong, compressed, rounded above; neck moderate; body rather full, slightly depressed. Feet short, stout, placed a little behind the centre of the body; legs bare a little above the joint; tarsus very short. Hind toe very small, with a narrow, free membrane.

Plumage dense, soft, and elastic; of the head and neck short, blended, and splendent; of the occiput and nape considerably elongated; of the other parts in general broad and rounded. Wings of moderate length, acute. Tail short, rounded, of fourteen acute feathers, of which the two middle extend five-twelfths of an inch beyond the next.

Bill grayish-black, tinged with yellow. Iris reddish-orange. Feet vermilion; claws dusky. Head and upper part of neck deep-green with purplish reflections, the top of the head of a darker tint, with less vivid gloss. A longitudinal band on the hind neck and the back, grayish-brown, the feathers edged with paler. The rump and upper tail-coverts greenish-black. Tail feathers grayish-brown, irregularly variegated and margined with grayish-white, that color enlarging on the outer feathers. Lower part of neck pure white; breast and middle part of abdomen dull purplish-chestnut. A large patch of white on each side of the rump, with a band of the same towards the tail. Lower tail-coverts greenish-black, with bright green and blue reflections; axillaries and lower wing-coverts pure white.

Length to end of tail, 20½ inches; extent of wings, 31½; weight, 1 pound 9 ounces.

FEMALE: Bill dull yellowish-green. Iris paler than in the male. Feet as in the male, but lighter. The upper parts are blackish-brown, the feathers edged with light reddish-brown; the throat and sides of the head are light reddish-brown, which is the prevailing color over the lower part of the neck, a portion of the breast, and the sides, of which, however, the feathers are margined with dusky; the middle of the breast white. Smaller wing-coverts dull brownish-gray; alula and primaries as in the male; inner secondaries brownish-black; the speculum as in the male, but paler, and changing to blue; the secondary coverts tipped with white. Tail nearly as in the male.

Length to end of tail, 17 inches; extent of wings, 29½; weight, 1 pound 1 ounce.

Spoonbills are seldom found in large numbers, but opportunities for shooting them will frequently occur when in the pursuit of other fowl. They associate with mallards when feeding. and their notes sound much alike. They fly very closely and irregularly together, and when a small flock comes to the decoys (for they decoy and come to the mallard call exceedingly well), the experienced hunter usually counts on securing fully half their number. A very slight blow brings them down. In fact, 1 have frequently seen whole flocks of five to eight individuals killed with both barrels. They are very poor divers, even when wounded, and try to escape by hiding, if possible. It is commonly easy to approach them within favorable distance, when feeding along the edge of the shore, by the exercise of ordinary caution; and when fired at, those unhurt spring at once perpendicularly into the air some ten or fifteen feet before taking a direct course away.

CHAPTER XXI.

DUSKY DUCK (ANAS OBSCURA).

Local name, "Black Duck."

ADULT MALE: Bill about the length of the head, higher than broad at the base, depressed and widened towards the end, rounded at the tip.

Head of moderate size, oblong, compressed; neck rather long and slender. Body full, depressed. Feet short, stout, placed a little behind the centre of the body; legs bare a little above the joint; tarsus short, a little compressed; hind toe extremely small.

Plumage dense, soft, and elastic; on the head and neck the feathers linear-oblong, on the other parts in general broad and rounded. Wings of moderate breadth and length, acute. Tail short, much rounded, of eighteen acute feathers, none of which are reserved.

Bill yellowish-green; the unguis dusky. Iris dark-brown. Feet orange-red; the webs dusky. The upper part of the head is glossy brownish-black, the feathers margined with light-brown;

221

the sides of the head and a band over the eye are light grayish-brown, with longitudinal dusky streaks; the middle of the neck is similar, but more dusky. The general color is blackish-brown, a little paler beneath, all the feathers margined with pale reddish-brown. The wing-coverts are grayish-dusky, with a faint tinge of green; the ends of the secondary coverts velvet-black. Primaries and their coverts blackish-brown, with the shafts brown; secondaries darker; the speculum is green, blue, violet, or amethyst-purple, according to the light in which it is viewed, bounded by velvet-black; the feathers also tipped with a narrow line of white. The whole under surface of the wing and the axillaries white.

Length to end of tail, 24½ inches; extent of wings, 38½; weight, 3 pounds.

ADULT FEMALE: The female, which is somewhat smaller, resembles the male in color, but is more brown, and has the speculum of the same tints, but without the white terminal line.

Length to end of tail, 22 inches; extent of wings, 34½.

In form and proportions the dusky duck is very closely allied to the mallard.

Though the black duck is not, strictly speaking, a Western fowl, yet it is quite frequently found there in company with mallards, and is generally known amongst Western sporting-men by the name of the black mallard. This is a mistaken idea; it is a separate and distinct species.

In certain parts of the Eastern States they are, in the proper season, very numerous, being, in fact, to that part of the country what the mallards are to the Western States, more numerous than any other shoal-water ducks, and most eagerly pursued by sportsmen. They are exceedingly wary, and take alarm at the slightest noise, springing at once perpendicularly into the air, often to a height of twenty feet, when they take a direct course at great speed. Their habits in the West are almost identical with those of the mallard, and their food the same.

CHAPTER XXII.

ADULT MALE: Bill longer than the head, higher than broad at the base, depressed, and a little widened towards the end, rounded at the tip. Upper mandible with the dorsal line sloping; the ridge very broad at the base, with a large depression, narrowed between the nostrils, convex towards the end; the sides nearly erect at the base, gradually becoming more horizontal and convex towards the end.

Head of moderate size, oblong, compressed; neck extremely long and slender; body very large, compact, depressed. Feet short, stout, placed a little behind the centre of the body; legs bare an inch and a half above the joint; tarsus short, a little compressed, covered all round with angular scales, of which the posterior are extremely small. Hind toe extremely small, with a very narrow membrane.

A portion of the forehead about half an inch in

length, and the space intervening between the bill and the eye, are bare. Plumage dense, soft, and elastic; on the head and neck the feathers oblong, acuminate; on the other parts in general broadly ovate and rounded; on the back short and compact. Wings long and broad; the anterior protuberance of the first phalangeal bone very prominent; primaries curved, stiff, tapering to an obtuse point, the second longest, exceeding the first by half an inch, and the third by a quarter of an inch; secondaries very broad and rounded, some of the inner rather pointed. Tail very short, graduated, of twenty-four stiffish, moderately broad, pointed feathers, of which the middle exceeds the lateral by two inches and a quarter.

Bill and feet black, the outer edges of the lower mandible and the inside of the mouth yellowish flesh-color. The plumage is pure white, excepting the upper part of the head, which varies from brownish red to white, apparently without reference to age or sex, as in *cygnus Americanus* and *anser hyperboreus*.

Length to end of tail, 68 inches; bill along the ridge, $4\frac{7}{12}$; from the eye to the tip, 6.

Young after the first moult:

In winter the young has the bill black, with

the middle portion of the ridge, to the length of
an inch and a half, light flesh-color, and a large,
elongated patch of light dull purple on each side;
the edge of the lower mandible and the tongue
dull, yellowish flesh-color. The eye is dark-brown.
The feet dull yellowish-brown, tinged with olive;
the claws brownish-black, the webs blackish-brown.
The upper part of the head and the cheeks are
light reddish-brown, each feather having towards
its extremity a small, oblong, whitish spot, narrowly margined with dusky; the throat nearly
white, as well as the edge of the lower eyelid.
The general color of the other parts is grayish-
white, slightly tinged with yellow; the upper part
of the neck marked with spots similar to those
on the head.

Length to end of tail, 52½ inches; extent of
wings, 91; weight, 19 pounds 8 ounces, the bird
very poor.

The swan is the largest, most spotless, and
most elegantly formed of all wild fowl. No other
gives the sportsman so much pleasure to secure;
and as their general scarcity and extreme wari-
ness render their capture comparatively rare, especial pride and honor are attached to the event.
Though there are several different varieties in-

cluded in the fauna of our country, their habits
are mainly alike, and I shall therefore deal only
with that variety I am most familiar with—the
cygnus buccinator, or trumpeter swan, the largest
of its kind, and most common to the valley of
the Mississippi.

The trumpeter swan first makes its appearance
from the north just before the first severe frosts,
and resorts during the day to the large open
bodies of water where it may sit far enough
from shore to feel secure from its numerous ene-
mies. In the early evening it either swims in
to the shoal water along the edges to feed, or
takes flight to some neighboring shallow pond or
slough for the same purpose. Its food, which it
never entirely submerges the body to obtain, consists of the roots, leaves, and seeds of different
vegetables (particularly the roots of pond-weed),
aquatic insects, small reptiles, and land-snails.
" Often it resorts to the land, and there picks
at the herbage, not sideways, as geese do, but
more in the manner of ducks and poultry."
Their flight is commonly in form similar to that
of the wild goose, though much faster and well
elevated, particularly when over the laud. When
alarmed, they are unable to rise or turn side-

ways at all suddenly. I once knew two swans
of separate flocks, coming from opposite directions,
to fly against each other, when one was so badly
hurt that he fell to the water, and was afterwards captured by my hunting companion, a man
named Tyler. Tyler had seen one flock approaching, and, as they came over his head, had fired
and killed one of their number, when almost immediately the collision occurred with a flock which
he had hitherto unnoticed, and one fell, as before
stated. The confusion incident to his shooting
was no doubt the immediate cause of this rather
remarkable accident.

Swans invariably rise against the wind, and if
it be calm they are unable to lift themselves
above the water before flapping along upon its
surface many yards, during which the strokes of
their wings against the water produce a rapid
succession of loud crackling noises, which may be
distinctly heard a long distance. Advantage is
taken by the skilful sportsman of this habit of
rising against the wind, and if he can find them
sitting to leeward in some narrow river or slough,
whose banks are bordered with high timber,
he is almost certain to capture some of their
number. Paddling cautiously down-wind to-

wards them along the shores and behind the
points, until as near as possible without exciting
observation, if still too far, he strikes out boldly
in line for them, urging his boat at its utmost
speed, and, though working his best at the paddle,
being careful to crouch low and avoid all unnecessary movements. At sight of him the swans
do not usually rise immediately, but sit turning
about · perplexedly for a few seconds, as though
conscious of their danger, yet at a loss how to
escape from it; and finally, as their only expe-
dient—desperate, though preferable to remaining
where they are—they are compelled to fly towards him, with faint hopes, by keeping close to
the further shore, of passing in safety by. These
hopes, however, to their sorrow and the sportsman's joy, if he thoroughly understands his business and follows it, are seldom fully realized.
His course, after they start from the water, is
simply to meet or head them, and when they
have arrived sufficiently near to choose his birds
and kill them. If he wishes one for food, a
cygnet (gray one) should be selected, for the
flesh of the young swan, though coarse, is tender
and exceedingly rich in flavor, while that of the
older ones is more tough and unsavory; but if

the skin is the principal part wanted, the larger
and older the bird the better.

Another very common method of hunting swans
is to take a stand where they pass and repass
from one pond to another. The same route is
nearly always taken, and during hard head-winds
they frequently fly quite low. They are very often,
too, driven by a strong side-wind quite near the
various points along the shores of the lakes which
they frequent, and many are thus killed. They
may also be approached, by using ordinary caution, in the sculling-float.

Single swans of this variety may frequently be
turned from their course by imitating their notes,
which resemble greatly those of a trumpet; and
because of this peculiarity of note the name
"trumpeter" was given them.

Never shoot at swans when the breasts are presented. Wait until they have slightly passed you;
and if your gun is loaded with shot smaller than
No. 1, aim to strike them in the head or neck.
It is useless to shoot small shot at the body, where
the covering is so very thick. If the outside feathers be plucked off carefully, a most beautiful
coat of snow-white down will be found underneath, fully an inch in thickness, and excelling in

silky fineness all other fur. The skins with this down remaining upon them form one of the chief articles of export of the Hudson's Bay Fur Company, and are used by the wealthy chiefly for trimming outside winter garments. Skins of birds killed in spring are more valuable than those of fall birds, which are usually " pin-feathery."

CHAPTER XXIII.

Commonly known as "The Wild Goose."

Adult male : Bill shorter than the head, rather higher than broad at the base, somewhat conical, depressed towards the end, rounded at the tip.

Head small, oblong, compressed. Neck long and slender. Body full, slightly depressed. Feet short, stout, placed behind the centre of the body; legs bare a little above the tibio-tarsal joint; tarsus short, a little compressed, covered all round with angular, reticulated scales, which are smaller behind; hind toe very small, with a narrow membrane. Wings of moderate length, with an obtuse protuberance at the flexure.

Plumage close, rather short, compact above, blended on the neck and lower parts of the body. The feathers of the head and neck very narrow; of the back very broad and abrupt; of the breast and belly broadly rounded. Wings when closed extending to about an inch from

236

the tail, acute. Tail very short, rounded, of eighteen stiff, rounded, but acuminate feathers.

Bill, feet, and claws black. Iris chestnut-brown. Head and two upper thirds of the neck glossy-black. Forehead, cheeks, and chin tinged with brown. .Lower eyelid white ; a broad band of the same across the throat to behind the eyes. Rump and tail feathers also black. The general color of the rest of the upper parts is grayish-brown; the wing-coverts shaded into ash-gray; all the feathers terminally edged with very pale brown; the lower part of the neck passing into grayish-white, which is the general color of the lower parts, with the exception of the abdomen, which is pure white, the sides, which are pale brownish-gray, the feathers tipped with white, and the lower wing-coverts, which are also pale brownish-gray. The margins of the rump and the upper tail-coverts pure white.

In very old males I have found the breast of a fine pale buff.

Length to end of tail, 43 inches ; extent of wings, 65 ; weight, 7 pounds.

ADULT FEMALE: The female is somewhat smaller than the male, but similar in coloring, although the tints are duller. The white of the throat is tinged with brown; the lower parts are always more gray, and the black of the head, neck, rump, and tail is shaded with brown.

Length, 41 inches; weight, 5¾ pounds.

No one species of the whole family of wild fowl is so familiar to the inhabitants of our country as the wild goose, for they are to be met with or seen upon their migrations in nearly every State in our Union. Their form of flight is generally in a straight line, with perhaps a second one branching from the main one, thus: * * * * * * , in either case led by an old gander, who, after acting as guide and breaking the way, as it were, through the air for a time, drops back, and is relieved by another, who also is relieved in turn. "In foggy weather or during severe snow-storms they frequently appear to get bewildered, and act as though they had lost their way. On such occasions they often alight to rest and recollect themselves." Before alighting from a long journey in a strange place, they always commence cackling and honking loudly, as though discussing the safety or advantages of the place. They feed upon grass, grain, pond-lily nuts, insects, and the roots of a peculiar plant which grows in shallow, stagnant waters,

and which is called by the hunters goose-flag.
When feeding, an old gander is usually placed
on guard, who warns them by an expressive
honk of any threatening danger. They never
dive for food, but feed when on the water like
the shoal-water ducks, by immersing the head
and neck.

On the bars or flats of the Mississippi, where
they often resort in great numbers for sand to
aid in digesting their food, boxes are sunk near
the edge of the water, in which the hunter lies
in wait for them. He seldom fires, however,
unless they are over the land, as when they fall
into the water the swift current carries them so
far that the loss of much time and labor is occasioned in recovering them. They are so heavy
to carry, and bring so poor a price—usually
about seventy-five cents each—besides being very
wary, thus making a decent remuneration for
pursuing them rather uncertain, that the market-
hunter seldom makes it an especial business,
unless at a time when ducks are scarce. For
the benefit of the novice I will describe some
of the methods employed in their capture, with
a few rules and hints..

One of the best plans with those who understand it is to use the sculling-float; the novice, however, will need to be pretty thoroughly trained before he can hope to be very fortunate, for sculling upon wild fowl successfully is almost a science in itself. One must thoroughly understand their habits, and be able to determine by their various actions or talk the state of their suspicions, and so govern his approaches accordingly. I will give a few of the principal rules, however, which will be necessary for the novice to follow. Some minor ones, which it is almost impossible to describe here, I will leave him to learn from experience.

We will suppose him, fully equipped in his float, to have discovered a flock of geese sitting in a position favorable for his approach. His guns (two or three should be used in this branch of the sport) are properly loaded, and his float trimmed, according to the season and the nature of the locality, with flags, rice, or brush, as he may deem least liable to excite suspicion; or, if in winter, when masses of ice are floating about, by a cake of ice laid judiciously across the bow. He must now, if not already there, get to the windward of the geese without their observing him; then, taking care not to rock his boat or make the least noise, proceed to scull down upon them

almost directly with the wind. By their actions
he must regulate his approach, whether fast or
slow. They may often huddle together, seemingly
to discuss the nature of the object approaching
them and the prudence of remaining; and then,
as though considering their fears groundless, may
spread apart again. If the hunter now uses due
caution, he will seldom fail to get within reasonable gunshot; but on their huddling close together, stretching their necks, and turning about
head to wind, he may be assured they intend to
rise, and if he is within distance, as soon as they
turn he must be ready to fire. Turning the bow
slightly by a quick stroke of the oar to favor his
awkward position for shooting, and grasping his
heaviest gun, he rises quickly, and pours in both
barrels as rapidly as possible, yet taking time to
select the thickest portion of the flock, and to
deliver his loads with most killing effect. Then
instantly seizing his second gun, he is to secure
as many as possible as they fly by or turn from
him. If he is possessed of a third gun, he will first
kill his most active cripples; then, after securing
the wounded, he may gather the dead. It is a
prevailing idea amongst sportsmen that geese,
and in fact all wild fowl, should always be ap-.

proached from the leeward to guard against their
exquisite sense of smell. This the wild-fowler
never need to bother himself about. Their sense
of smell may be exceedingly acute, but I doubt
very much their ability to recognize danger by
such means. They possess, however, a remarkable sense of hearing, and often take alarm at
the least unusual sound; for this reason one can
approach with less caution from the leeward.
This raises the question, Why, if we need to use
more caution, do we approach from the windward in our sculling-float? Simply because they
are obliged to rise against the wind. If we were
to approach from the leeward, they would naturally swim away from us, and would give us
little or no warning when they intended to rise,
besides seldom allowing us to get as near as
though we approached from the opposite side,
being loath to fly towards the object of their suspicions until they are obliged to. Another thing:
besides our poorer chances of a family shot with
our first gun, our opportunity for using the
second is entirely lost. I would particularly impress on the mind of the novice the importance
of looking into the whys and wherefores and
studying the reasons of things pertaining to wild-

fowl shooting; he will thereby learn to avoid difficulties, and to take advantage of opportunities which he might otherwise overlook. His progress in the art will be much more rapid and his success more certain.

Geese are particularly fond of young rye and winter wheat, and in the Western country where these are cultivated to any great extent good shooting may often be had. When the ponds begin to freeze over in the early winter is the best season. The hunter usually selects a position in a corner of some one of the numerous rail-fences, or amongst the high "horse-weeds" bordering the field, and, after setting out his decoys (if he is fortunate enough to be possessed of any), patiently awaits their coming. On their approach he commences to call, and seeing his decoys they come down fearlessly: those he kills he props up with sticks to decoy others. If he chances to secure a wing-broken one, he ties him to a stake amongst his other decoys, and when other geese come in sight the poor cripple never fails to call loudly. A box sunk near the centre of the field is better, if it can be fixed conveniently, than a blind near the edges, for the geese are inclined to avoid the fences as much as possible. At this time of the

year the air-holes in the river, where they come
to get water and to roost at night, are often better
places than the fields for securing a good bag.
They come to the decoys there with less fear or
suspicion. When wounded, geese dive easily, and
often swim long distances under water. Opportunities for shooting them will frequently occur
when in the pursuit of ducks. No. 1 (Chicago) I
consider the best-sized shot for goose shooting,
though B or B B may with propriety be used in
large or very close-shooting guns.

l am astonished that Audubon has ascribed so
little weight to the wild goose. l have seldom
seen one weighing less than eight or nine pounds,
and have seen many over twelve; and one which
l killed myself, the largest l ever saw, weighed
eighteen pounds.

SHOOTING GEESE FROM SCULLING FLOAT.

CHAPTER XXIV.

The "Brant," amongst Western sportsmen.

ADULT MALE: Bill shorter than the head, much higher than broad at the base, somewhat conical, depressed towards the end, rounded at the tip.

Head of moderate size, oblong, compressed. Neck rather long and slender. Body full, slightly depressed. Feet rather short, strong, placed rather behind the centre of the body; legs bare a little above the joint; tarsus rather short, a little compressed, covered all round with angular reticulated scales, which are smaller behind.

Plumage close, full, compact above, blended on the neck and lower part of the body, very short on the head. Feathers of the head and neck very narrow, on the latter part disposed in oblique series separated by grooves; of the back very broad and abrupt; of the breast and belly broadly rounded. Wings rather long and broad. Tail very short, rounded, of sixteen broad, rounded feathers.

241

Bill carmine-red, the unguis of both mandibles white. Edges of eyelids dull orange. Iris hazel. Feet orange, webs lighter, claws white. Head and neck rich grayish-brown, the upper part of the former darker; a white band margined with blackish-brown on the anterior part of the forehead along the bill. The general color of the back is deep-gray, the feathers of its fore part broadly tipped with grayish-brown, the rest with grayish-white; the hind part of the back pure deep-gray. Wings grayish-brown, but towards the edge ash-gray, as are the primary coverts and outer webs of the primaries; the rest of the primaries and the secondaries are grayish-black, the latter with a narrow edge of grayish-white, the former edged and tipped with white. Breast, abdomen, lower tail-coverts, sides of the rump, and upper tail-coverts white, but the breast and sides patched with brownish-black; on the latter intermixed with grayish-brown feathers.

Length to end of tail, 27¼ inches; weight, 5¼ pounds.

ADULT FEMALE: The female, which is somewhat smaller, resembles the male. The white margins of the wing-feathers not so distinct. Weight, 4 pounds 4 ounces.

By Audubon the white-fronted goose (*anser albifrons*) and the snow-goose (*anser hyperboreus*) have been classified and described (and I think properly so) as two entirely distinct and separate species ; though many naturalists, seeing in them many striking points of resemblance, and being confused by their varying appearances, produced by age and change of season, coupled with their lack of familiarity with them whilst breeding, have adjudged them to be identical in character, of but one species, and differing only in degree of maturity. The opinion of Audubon is, I believe, that most generally sanctioned by Western sportsmen, amongst whom, however, both species receive the general appellation of " brant," the white-fronted being familiarly known as the harlequin variety, from the irregular, patchy coloring of their breast-feathers, while the snow-geese are all called fish-brant, and as such are never pursued for the table. The younger ones of this latter species are further characteristically distinguished as bald brant or white-heads.

To the peculiar habits of the snow-goose I have paid but little attention, being rather a poor naturalist, and led to consider them by my hunting companions as unworthy of pursuit. I have frequently, however, seen them feeding upon grain, and observing that Audubon and several others have pronounced them excellent eating, am inclined to think the prevailing idea amongst hunters in regard to the fishy flavor of their flesh to be poorly founded.

The flesh of the white-fronted goose is, I think, acknowledged by all who have partaken of it to be delicate and well-flavored.   They feed almost entirely by day, and chiefly upon grass, grain, and other vegetable matter, which they procure upon the low, wet prairies and grain-fields; leaving for the purpose, early in the morning, the large lakes and ponds where they roost at night, and returning to them again in the afternoon about sundown.   On these journeys, which are often several miles in length, they fly high in air, much in the manner of wild geese (whose wing-strokes, however, are not quite so fast), and seldom lower their flight until directly over their feeding-grounds or roosting-ponds, when all dart down together in a confused, zigzag, and irregular manner, cackling loudly, and uttering the while most discordant sounds. (This habit is also characteristic of the snow-geese, and perhaps in a greater degree ; their notes, however, are widely different.) When their feeding-grounds and roosting-places are near together, they may fly quite low, particularly if against a strong head-wind, and good sport may then be had on the passways. Single ones may frequently be called within gun-shot, when coming in to roost (though never when in the fields), by imitating their call-notes correctly—an achievement, however, rather difficult to most persons, as they are pitched in so high a key. A slight resemblance to the note may be expressed by the sounds *ela, eleck, ela, ela, eleck.* They are exceedingly acute, sharp-sighted birds, and can " climb " out of gun-shot (as the hunters term their flying upwards when frightened at the appearance of danger beneath them) faster than most wild-fowl.

In the corn-fields they are frequently shot from holes in the ground or from blinds built of corn-stalks in a manner similar to that described in the foregoing chapter. No. 3 shot, Chicago size, is sufficiently large for brant shooting, as these birds are not very tenacious of life. They are rarely found along the Eastern coast.

CHAPTER XXV.

ADULT MALE : Bill as long as the head, deeper
than broad at the base, the margins parallel,
slightly dilated towards the end, which is rounded,
the frontal angles rather narrow and pointed.

Head rather large, compressed, convex above.
Eyes small. Neck of moderate length, rather
thick. Body full, depressed. Wings small. Feet
very short, strong, placed rather far behind ; tarsus very short, compressed, anteriorly with narrow
scutella continuous with those of the middle toe,
and having another series commencing half-way
down and continuous with those of the outer toe,
the rest reticulated with angular scales. Hind
toe small, with an inner expanded margin or web ;
middle toe nearly double the length of the tarsus, outer a little shorter. Claws small, compressed, that of the first toe very small and

246

curved, of the third toe larger and more expanded than the rest.

Plumage dense, soft, blended. Feathers of the upper part of the head small and rather compact, of the rest of the head and neck small, blended, and glossy. Wings shortish, narrow, pointed; primary quills strong, tapering, the first longest, the second almost as long, the rest rapidly diminishing; secondary quills broad and rounded, the inner elongated and tapering. Tail very short, much rounded or wedge-shaped, of fourteen feathers.

Bill black, with a tinge of green. Iris bright carmine. Upper part of the head and a space along the base of the bill dusky; a small, transverse band of white on what is called the chin; the rest of the head and the neck all round for more than half its length of a rich brownish-red. A broad belt of brownish-black occupies the lower part of the neck and the fore part of the body, of which the posterior part is the same color, more extended on the back than under the tail. Back and scapulars white or grayish-white, very minutely traversed by undulating black lines; wing-coverts similar, but darker. Alular feathers grayish-brown. Primary quills brownish-black,

tinged with gray towards the base; the shaft
brown. Secondaries ash-gray, whitish, and undulated with dark-gray towards the end; five of
them having also a narrow stripe of black along
their outer margin. Tail brownish-gray, towards
the end ash-gray. The lower parts white, the
sides and abdomen marked with fine undulating
gray lines, of which there are faint traces on most
of the other feathers. The feet are grayish-blue,
tinged with yellow.

Length to end of tail, 22 inches; extent of
wings, 33; weight, 3¾ pounds.

Adult female: The female has the bill colored
as in the male; the iris reddish-brown; the feet
lead-gray; the upper parts grayish-brown; the
top of the head darker, its anterior part light-
reddish; the chin whitish; the neck grayish-brown,
as are the sides and abdomen; the breast white;
wing-coverts brownish-gray; primary quills grayish-brown, dusky at the end; secondary quills
ash-gray, five of the inner with an external black
margin, the innermost grayish-brown like the back,
and with some of the scapulars faintly undulated
with darker. Tail grayish-brown, paler at the
end; axillars and smaller under wing-coverts white,
as in the male.

Length to end of tail, 20¼ inches; extent of wings, 30¾; weight, 2¾ pounds.

No one species of the whole duck tribe so sorely puzzles the uninitiated to secure as the wary, gamey, and highly-prized canvas-back. The mallard and the various other shoal-water ducks he frequently finds opportunities for " bush-whacking " along . the shallow edges of the ponds and sloughs where there may be sufficient cover; or he may catch them flying low down over the narrow creeks or bushy points, where, if he is sufficiently skilful, he may also manage to secure a brace or two. With the deep-water varieties, and the canvas-back most especially, the case is decidedly different. They are too well contented to sit tantalizingly out of reach near the middle of the open waters, and rarely venture into the smaller ponds, where they would be obliged to sit in too close proximity to the shores, knowing far too well the danger to be apprehended from such a proceeding. The art here, then, is not simply to aim straight and pull the trigger at the proper moment, but it is also to know how to approach them, or induce them to approach you within gunshot, with least loss of time and labor.

From the circumstance of the lake-shores being

almost invariably inundated in the spring or covered with high weeds and brush, thus leaving no bare ground upon which to operate the system of toling canvas-backs, as practised upon the Chesapeake Bay, is never made use of in the West. As it may, however, prove of interest to my readers to know how the sport is conducted, I will quote Dr. Sharpless's vivid description, never myself having had the pleasure of witnessing it:

"A spot is usually selected where the birds have not been much disturbed, and where they feed at three or four hundred yards from, and can approach to within forty or fifty yards of, the shore, as they will never come nearer than they can swim freely. The higher the tides and the calmer the day the better, for they feed closer to the shores; and see more distinctly. Most persons on these waters have a race of small, white or liver-colored dogs, which they familiarly call the toler breed, but which appear to be the ordinary poodle. These dogs are extremely playful, and are taught to run up and down the shore, in sight of the ducks, either by the motion of the hand or by throwing chips from side to side. They soon become perfectly acquainted with their business, and, as they discover the ducks approaching them, make their jumps less high, till they almost crawl on the ground to prevent the birds discovering what the object of their curiosity may be. This disposition to examine rarieties has been taken advantage of by using a red or black handkerchief by day and a white one by night in toling, or even by gently plashing the water on the shore. The nearest ducks soon notice the strange appearance, raise their heads, gaze intently for a moment, and then start for the shore, followed by the rest. On many occasions I have seen thousands of them swimming in a solid mass direct to the object; and by removing the dog further into the grass they have been brought within fifteen feet of the bank. When they have approached to about thirty or forty yards, their curiosity is generally satisfied, and, after swimming up and down for a few seconds, they retrograde to their former station. The moment to shoot is while they present their sides, and forty or fifty ducks have often been killed by a small gun. The black-heads tole the most readily, then the red-heads, next the canvas-backs, and the bald-pates rarely. To prevent the dogs, whilst toling, from running in, they are not allowed to go into the water to bring out the ducks, but another breed of large dogs, of the Newfoundland and water-spaniel mixture, are employed."

The usual method of taking canvas-back in the West is by the aid of decoys, shooting either from a sink-box—a battery built of brush, etc., on a paddle-boat—or from a blind built in some favorable position along the edges of the willows. The first named is but little used, however, on account of the frequent difficulty of conveying it from one lake to another, part of the distance perhaps being through thick brush or willows or across dry ridges. Then, as it necessitates the services of two men to work it, but few more ducks can be killed from it than by other methods where each man may hunt separately.

When ducks get bedded (*i.e.*, in the habit of sitting in large bodies in the same place for purposes of feeding or otherwise) in the large, open waters, and act shy of decoys placed near the shore, a common paddle-boat covered with brush and weeds answers nearly as well as the sink-box, and costs much less labor to prepare; and, when done with, the brush may be thrown off, and the labor of towing about the "sink" avoided. Some of my readers may wish to try the sink, however, so I will give a brief description of its build and the manner of using it.

The box in which the shooter lies should be of pine, sides and bottom one inch and ends two inches thick, and of proportions adapted to the size of the person to occupy it; six feet long, two feet wide, and thirteen inches deep being proper for an ordinary-sized man. Along each side and across the ends, one inch below the top edge of the box, two-by-four-inch pine timbers are fastened, framed together to equal height, and extending on all sides two and a half feet from the box. This frame should be slanted off on top fully an inch towards the ends to give a pitch to the deck, and on the under side should also be reduced in the same manner to make it as light as possible for handling. The frame is next covered with a pine platform a half-inch thick, which is further strengthened by the addition of a brace reaching from the centre of the box on each side. This platform is bounded on the three sides by hinged wings of cotton-cloth, which are two feet wide, fastened to a pine frame-work, and so constructed as to admit of being folded back upon the platform when not in use. At the fourth side or head of the sink the wing, instead of

being made entirely of cloth, is partly composed
of two half-inch pine boards, eight inches
wide, hinged together, and extending the width
of the platform, to which the inner board is
fastened by strong hinges ; the rest of the wing,
which is equal in width to the others, is of
cloth, and all the wings are joined together by
angle-pieces of the same material. A border of
sheet-lead, three inches in height, is to be tacked
completely around the outside edge of the box,
and inclined outwards, as the flare of a boat, to
throw off any little ripple that might otherwise
wash into the box. Across the head, and about
half-way round the sides, where the tendency of
the waves to wash in is always greatest, a
second circular rim of lead four inches high, as
a double precaution, should also be fastened and
flared like the other. This outside rim should be
placed about fifteen inches from the end of the
box. A short rope, about six feet in length, is
fastened at each end, about three feet apart, to
the cross-timber at the head of the box, to the
middle of which rope the anchor-line is attached.
A second anchor is also sometimes used, which
should be fastened to the foot of the platform.
This, however, except in very shallow water, is

needless. To finish, the whole thing is now to be painted as near the color of the water as possible, and when dry is ready for use.

A bed of hay or straw is prepared in the bottom, on which the shooter is to lie, a pillow placed at the end for his head, and the sink is next towed out and anchored in the desired position. The decoys are next set out, the guns and ammunition transferred from the paddle-boat, and after adding sufficient ballast with the weight of the shooter to sink the edge of the platform to the surface of the water, the shooter takes his place, and his companion leaves him. The companion's duty now is to rout up the ducks occasionally when they get settled; to secure the cripples, if possible; to pick up the dead; and to release the shooter when necessary.

From the position of the shooter it is evident he can only shoot in very limited directions; the decoys must therefore be so arranged that ducks coming to them will approach in such a manner as shall be most favorable to his condition for shooting. The arrangement fulfilling this requirement most perfectly is as follows: not less than one hundred decoys should be used, placed square to the right fifteen yards, to

the left twenty-five yards, from this line narrowing gradually to a point about ten yards to the
left of a direct line to leeward, and at a distance
of thirty-five yards from the sink; from this
point, three or four tolers, ten to fifteen yards
apart, to leeward, and inclined towards the direction
the ducks mainly approach from or pass by. Near
the centre of this triangle, which is the figure
the flock now represents, the decoys should be
scattered a trifle more than at other places, and
the ducks will endeavor to alight there. A
few decoys should be fastened to the platform
of the sink.

This arrangement of decoys is the one most suitable for sink-box shooting. No matter on which
side ducks may be, when they observe the decoys
they almost invariably approach to alight, against
the wind, if it be blowing at all; and as the decoys are now placed, they will come in over or
very close to the leeward point of the triangle,
because in so approaching that point is nearest to
them, and they seldom take a roundabout course
without a reason for it. More decoys are set to
the left of the sink, because it is much easier for
the shooter to swing his gun on that side than
towards the right, as he might be obliged to do

if there were equal numbers on each side, when the ducks would be as likely to turn one way as the other. If during the day the wind should change, it will be necessary to alter the position of the battery. This may be done without taking up the decoys, if they are arranged as directed; all that is necessary being to shift the box to windward.

When ducks are flying by, especially on very calm days, a good way to attract their attention to the decoys is to raise your hands above the edges of the box and wave them quickly to and fro, imitating as near as may be the action of ducks when flapping their wings or when alighting on the water. This turning involuntarily the eyes of the passing ducks towards the decoys, they come in readily.

The shooter should be careful not to rise up too soon when ducks are approaching. Wait until they are over the "tail" decoys, and if there is a large flock, and they choose to alight, let them do so; and when you catch a sufficient number together, rise and fire quickly. A second gun often adds considerably to the score, if the shooter can handle it rapidly. A strap fastened across the top of one or both feet will help him in rising.

The method of shooting canvas-back from the

paddle-boat covered with brush is so similar to sink-box shooting, that it would simply be a waste of space to describe it; and the covering of the boat is so simple a matter that I will trust to the ingenuity of my readers without boring them with tedious explanations. I will just remark the position of shooting from the paddle-boat had better be from the knees than from a sitting position; bending forward when ducks may be approaching, instead of lying at length.

Canvas-back generally feed quite near the edges of the willows, as their food, which consists mainly of the bulbous roots of a certain water-grass, does not grow in the deeper waters, but rather in those portions of the lakes that are left bare during the summer and early fall months, when the water is generally low. During the fall, unless the water is unusually high, they are rarely seen; but in spring, when the melting of the winter's snow and ice, and the heavy rains, have raised the river and inundated these grassy plats, they may be often seen in flocks of thousands. When much disturbed, they feed mostly by night or early morning and evening, and sit during the middle of the day near the middle of the open waters opposite to their feeding-grounds. Before they become too wary, at intervals during the day small parties, as though unable to withstand the temptation any longer, get up and fly into these feeding-grounds, when, if the sportsman is prepared for them with plenty of decoys, he may have excellent sport and secure a goodly number ; but as it grows later in the season, they learn better, and come in less frequently, content to wait until evening, when, instead of taking to wing and flying in, the whole body swim in cautiously together. The blind for shooting in during the day should be selected with care *where they wish to feed*—not to one side or the other, or, instead of coming in to the decoys, many may pass by, and drop down out of range where they have found by experience their food grows in greater profusion.

As to the building of the blind and the arrangement of the decoys, both have been fully explained in the chapters especially devoted to those subjects, so I will not repeat them here. I shall, however, add a few hints as to the peculiar habits of the ducks when approaching decoys and at various other times, and will endeavor to explain how to best apply a knowledge of these habits to practical and profitable use.

Canvas-backs never set their wings and drop back from a height nearly perpendicular, as mallards and some other shoal-water ducks often do; but, when intending to alight, always lower as they approach, and if not sufficiently low when first over the decoys, sheer off and circle back again. They also frequently pass over or by the decoys low down, seemingly to a novice as though not seeing them, when, after going perhaps a hundred yards or more, if everything appears correct to them, they will turn about suddenly, and come in as though intending to stay. The experienced hunter, who understands this habit, also knows from their actions whether they will probably turn back again, or go on because of having seen something to alarm them, and so he either reserves his fire until they come the second time, or else improves his poorer, though only, chance as they first pass by. Through ignorance of this, many wild shots are coaxed from the novice which, if held back for a moment, might be turned to much better account.

Just before they get ready to alight, raise your head and shoulders slowly above the blind, and with the butt of the gun to the shoulder, and the muzzle just under the flock, be prepared to take advantage on the instant of their bunching or crossing. They will take no notice of you whatever, if you rise slowly and do not attract their attention by the suddenness of your movements, their eyes at the time being entirely occupied with the decoys or selecting a particular spot in which to light. If the flock is large, it may be advisable to allow them to do this, and when a sufficient number swim together, fire away. Frequently at the report of the gun, if they do not see the shooter, those unhurt may jump up and immediately drop down again, as though thinking there was no need of leaving while their companions, the decoys, appeared so quiet and contented; and if you are quick-motioned, you may often reload your breech-loader and secure more before they discover their mistake. This I have done repeatedly.

On rainy days they appear very uneasy, flying about continually, and dart to the decoys readily. It is on these days the big counts are usually made.

Cripples should be "shot over again" as soon as possible, and be careful not to let them see

you before firing, if you can help it. When
first wounded, they usually sit with head erect,
looking for the cause of their misfortune, and
are then easily killed ; but as soon as they discover the shooter, they dive, and, if only winged,
it is useless to follow them, unless, however, the
water is perfectly calm, when the ripple occasioned by their coming to the surface can be readily
seen. Even then it takes hard work and a long
chase to secure them, and if ducks are flying
well it is better to let them go. If struck in
the body, they may be more easily tired, and
then captured. They are exceedingly expert divers,
and can swim under water to much longer distances than any others of the vegetable-eating
ducks.

When chasing cripples, do not allow them to
remain long enough at the surface of the water
to regain their breath, if you can by shouting
prevent it, unless you are ready to shoot them
over again, which, by the way, you should always
be prepared to do, if desirous, before leaving
your blind. Their course under water you may
frequently follow by the minute air-bubbles escaping from them and coming to the surface
when they begin to get exhausted. If you are

prepared for their rising, and they come up sufficiently nigh, you may kill them by striking them across the head or neck with the edge of the paddle.

Frequently birds start off with the flock as though unhurt, and, after flying a few hundred yards, fall unnoticed dead or mortally wounded. It is advisable, therefore, to watch for some distance any you may think are struck, to see whether they may be recovered or not. They are exceedingly tenacious of life, and require hard hitting to secure them. No. 3 or 4 Chicagos are about the proper-sized shot to use for them.

Dogs are never used for retrieving canvasbacks in the West; the shooting is always done from a boat, when it would be inconvenient and unpleasant to have the dog continually getting in and out while wet and dripping with water; and as for their catching a crippled canvas-back, it is out of the question.

Canvas-backs are never known to breed along the Mississippi River or its tributaries, but betake themselves to some unknown regions of the far north where the white man never molests them. About the first of November, with their young, which are now almost fully grown, they take a direct course to the waters of the Chesapeake Bay and its confluent streams, rarely stopping upon the way, and here remain feeding upon their favorite food, the roots of the wild celery, until driven further south by the increasing cold weather. About Galveston Bay and the mouth of the Mississippi they are very plentiful, particularly if the weather on the Eastern coast has been very severe. In spring, instead of taking their roundabout course back north by the way of the Chesapeake, many choose the more direct route up the Mississippi, stopping here and there along the back waters of this river and its tributaries, where food may be found most abundant ; and in proportion to the severity of the previous winter at the East will they be found here in greater or less profusion.

CHAPTER XXVI.

RED-HEADED DUCK (ANAS FERINA).

ADULT MALE: Bill as long as the head, deeper than broad at the base; the margins parallel, slightly dilated towards the end, which is rounded; the frontal angles rather narrow and pointed.

Head rather large, compressed, convex above. Eyes small. Neck of moderate length, rather thick. Body full, depressed. Wings small. Feet very short, strong, placed rather far behind; tarsus very short, compressed anteriorly with narrow scutella, continuous with those of the middle toe, and having another series commencing half-way down, and continuous with those of the outer toe, the rest reticulated with angular scales; hind toe small, with an inner expanded margin or web; middle toe nearly double the length of the tarsus; outer, a little shorter. Claws small, compressed, that of the first toe very small and curved, of the third toe larger and more expanded than the rest.

Plumage dense, soft, blended. Feathers of the

upper part of the head small and rather compact, of the rest of the head and neck small, blended, and glossy. Wings shortish, narrow, pointed; primary quills strong, tapering, the first longest, the second almost as long, the rest rapidly diminishing; secondary quills broad and rounded, the inner elongated and tapering. Tail very short, much rounded or wedge-shaped, of fourteen feathers.

Bill light grayish-blue, with a broad band of black at the end, and a dusky patch anterior to the nostrils. Iris orange-yellow. Head and neck all around, for more than half its length, of a rich brownish-red, glossed with carmine above. A broad belt of brownish-black occupies the lower part of the neck and the fore part of the body, of which the posterior part is of the same color, more extended on the back than under the tail. Back and scapulars pale grayish-white, very minutely traversed by dark brownish-gray lines; the sides and abdomen similar, the undulations gradually fading away into the grayish-white of the middle of the breast; upper wing-coverts brownish-gray, the feathers faintly undulated with whitish towards the end. Primary quills brownish-gray, dusky along the outer web and at the end; secondaries ash-gray, narrowly tipped with white, the outer faintly tinged with yellow, and almost imperceptibly dotted with whitish; four or five of the inner of a poorer tint, tinged with blue, and having a narrow brownish-black line along the margin; the innermost like the scapulars, but more dusky. Tail brownish-gray, towards the end lighter. Axillar feathers and lower wing-coverts white. Feet dull grayish-blue, the webs dusky, the claws black.

Length to end of tail, 20 inches; extent of wings, 33; weight, 2½ pounds.

ADULT FEMALE: The female has the bill of a dusky bluish-gray, with a broad band of black at the end, and a narrow, transverse blue line, narrower than in the male. Iris yellow. Feet as in the male. The head and upper part of the neck dull reddish-brown, darker above and lighter on the fore part of the cheeks and along a streak behind the eye. The rest of the neck all round, and the upper parts in general, are dull grayish-brown, the feathers paler at their extremity; the flanks and fore part of the neck dull reddish-brown, the feathers broadly tipped with pale grayish-brown. The wings are as in the male, but of a darker tint and without undulations. The tail as in the male. Lower wing-coverts light-gray; those in the middle white, middle of breast grayish-white, hind part of abdomen light brownish-gray.

Length to end of tail, 21 inches; extent of wings, 32½; weight, 2 pounds 7 ounces.

Like those of the canvas-back, the habits of the red-head during the breeding-season are very poorly understood; in fact, the same may be said of all the deep-water ducks herein described, from the fact that none of them ever remain to breed in their winter quarters, but all take their departure in the spring to some secluded regions of the north, and in the fall return with their numerous progeny fully grown and well able to take care of themselves. Like the canvas-back, too, they are not very plenty in the West during the fall, unless the water is unusually high, but make their appearance in large numbers shortly after the opening of the rivers and lakes in spring.

Their food consists chiefly of the grass roots so much sought for by the canvas-backs, and also of corn, wild oats, and the seeds and blades of various water-grasses which grow along the shore in the fall, and which in spring are inundated. The botanical name of this grass, whose roots form its favorite food, I am unacquainted with. In shape the plant resembles the witch-grass, so common to old ploughed fields—blades long, thin, and rather narrower than the witch-grass, and its bulbs round, about the size of a pea, of a bright reddish-brown color on the outside, and on the inside a flaky white.

They assemble in large flocks on their feeding-grounds, and associate indiscriminately with both deep and shoal water ducks, but are rarely seen in flight with other than their own kind. They fly bunched closely together, but at regular rates of speed and in good order, and seldom pitch or dart about, as is the general habit of those shoal-water ducks that fly clustered together. They come in splendidly to decoys set out on their feeding-grounds, but very many flocks will pass them by closely without noticing them if on a passway. When they are coming to your decoys down-wind or with a side wind, rise just before they get to you, and as they double back to alight, "turn it loose" at the middle of the cluster. They are not very tenacious of life, and frequently six or eight may be killed at a shot in this way. At times they appear uncommonly foolish, returning to the decoys and lighting down immediately after being shot at. An imitation of their note, which much resembles the mewing

of a cat, will often help to attract their attention to decoys; but without decoys they will not
turn in for hearing it. They swim very fast
and are good divers, though not quite as expert
as the canvas-back; the methods of hunting
them are similar. In market they sell readily,
bringing, with the exception of canvas-backs,
a higher price than any other duck. They rise
from the water almost invariably against the
wind, and usually huddle before taking wing. St.
Louis fours or fives are the most killing shot
to use in shooting them.

CHAPTER XXVII.

SCAUP-DUCK (FULIGULA MARILA).

Local names, "Blue bill," "Broad-bill," "Black-head," and "Black-Jack."

ADULT MALE: Bill as long as the head, deeper than broad at the base, enlarged and flattened towards the end, which is rounded; the frontal angles narrow and pointed.

Head of moderate size. Eyes small. Neck of moderate length, rather thick. Body comparatively short, compact, and depressed. Wings small. Feet very short, strong, placed rather far behind; tarsus very short, compressed.

Plumage dense, soft, blended. Feathers of the head and neck short and velvety, those of the hind head a little elongated. Wings shortish, narrow, pointed; primary quills curved, strong, tapering, the first longest, the second very little shorter, the rest rapidly graduated; secondary broad and rounded, the inner elongated and tapering. Tail very short, much rounded, of fourteen feathers.

Bill light grayish-blue, the unguis blackish. Iris yellow. Feet grayish-blue, the webs and claws black. The head, the whole neck, and the fore part of the back and breast black; the head and neck glossed with purple and green, the rest tinged with brown. Hind part of the back, rump, abdomen, and upper and lower tail-coverts brownish-black. Middle of the back, scapulars, inner secondaries, anterior part of abdomen, and sides grayish-white, beautifully marked with undulating black lines. Middle of the breast white, wings light brownish-gray. Alula, primaries at the base and end, and the greater part of secondaries brownish-black; the speculum on the latter, white.

Length to end of tail, 16½ inches; extent of wings, 29; weight, 1 pound 6 ounces.

ADULT FEMALE: The female agrees with the male in the characters of the plumage and in the colors of the bare parts, but those of the former differ considerably. The head, neck, and fore part of the back and breast are umber-brown; and there is a broad patch of white along the fore part of the forehead. The upper parts in general are brownish-black, the middle of the back and scapulars undulated with whitish dots and bars. The primary quills are grayish in the middle, and the speculum is white, but of less extent than in the male. The greater part of the breast and abdomen is white; the sides and parts under the tail umber-brown.

Length, 16½ inches; extent of wings, 28; weight, 1 pound 6 ounces.

The males vary greatly in size, but in adult specimens there is little difference in coloring.

Blue-bill shooting, when they are to be found in abundance, as is frequently the case in spring, is one of the prettiest of sports; they come in to the decoys so readily, so often, and are off again with such amazing velocity, unless well held on, that the sportsman cannot fail of being pleased. Their flight, though rapid, is very steady, seldom high in air, excepting in long journeys over land, which they avoid as much as possible, and when in flocks frequently packed closely together, much in the manner of red-heads, thus affording excellent opportunities for killing several at a discharge. They die hard, and struggle to escape to the last, frequently, when shot in the air, diving the instant they strike the water, and coming up to its surface dead. When wing-broken, they swim under water to long distances, coming to the surface only for an instant to regain their breath, and diving again so quickly as to seldom allow time for shooting them over.

For food they depend chiefly upon wild rice and the bulbous grass-roots described as being also the favorite food of the canvas-backs and red-heads, and on their feeding-grounds all associate indiscriminately together. Blue-bills are also very partial to overflowed prairies and corn-fields, and are frequently to be found there in quite large numbers. They cannot spring at once into the air like many other ducks, but rise gradually as they go on, and get under good headway exceedingly quick. Unless the wind blows quite fresh, they may alight either with or against it, seldom turning back to alight if coming to the decoys down-wind. Though their flesh is well flavored, and generally in good order, they sell at very low prices, frequently at seventy-five cents per dozen in Chicago markets.

They are less cautious about approaching the shore than canvas-backs or red-heads, and large numbers are frequently killed over decoys from blinds built in the willows bordering some favorite feeding-ground. Almost any kind of a blind will do when shooting over decoys; only be careful not to attract their attention by any sudden or needless motions. Their note is a guttural, rolling sound, which may be slightly represented by the letters *krrr, krrr;* it is useless, however, to imitate it, excepting to turn their attention to decoys. They are usually abundant the entire spring. Chicago sixes are the proper-sized shot.

CHAPTER XXVIII.

Local names, "Ring-billed Duck," "Tufted Duck," and "Golden-
eye" (last very common, but erroneous).

ADULT MALE: Bill about the same length as
the head, rather deeper than broad at the base,
depressed and enlarged towards the end, the frontal
angles acute.

Head of moderate size. Neck rather long and
slender. Body full and depressed. Wings rather
small. Feet very short, strong, placed rather far
behind; tarsus very short.

Plumage dense, soft, blended, rather glossy.
Feathers of the middle of the head and upper
part of hind neck very narrow and a little elongated; the rest of the head and upper part
of the neck very short; of the back and lower
parts in general broad and rounded. Wings of
moderate length, narrow, acute. Tail very short,
rather broad, much rounded, of sixteen rounded
feathers.

Bill black; with a basal band, the edges of both mandibles and a band across the upper towards the end, pale blue. Iris yellow. Legs grayish-blue; the webs brownish-black. The head and upper part of the neck greenish-black, with purple reflections. A brownish-red collar or ring, broader before, on the middle of the neck. Its lower part all round, as well as the back, scapulars, smaller wing-coverts, and posterior part of the abdomen, brownish-black. Inner secondaries of the same color; outer bluish-gray on the outer web, light-brown on the inner, as are the primaries, of which the outer webs and tips are dark-brown. Tail brownish-gray. Chin white. Breast grayish-white. Sides and fore part of the abdomen grayish-white, minutely undulated with grayish-brown.

Length to end of tail, 18 inches; extent of wings, 28.

ADULT FEMALE: The female has the neck umber-brown; the upper part of the head darker; the back blackish-brown; the speculum darkish-gray, as in the male; the breast brownish-white; the loral spaces and chin pale-brown; the abdomen umber-brown.

Length, 16 inches.

The female of the ring-necked and scaup

ducks, which are alike in general color, differ in the speculum and in the peculiar form of the bill.

Ring-necked ducks, which are very similar in appearance and habits to blue-bills, are seldom found in very large numbers, though occasionally fair sport may be had with them. Where the water has flowed back amongst the thin, low willows which border the numerous lakes and sloughs, are their favorite resorts; though they may also be found quite frequently upon the overflowed prairies and corn-fields, associated with the blue-bills. The methods of hunting them are nearly identical, though decoys are of less advantage in their pursuit than in that of other deep-water ducks. They fly faster than most wild fowl; and, when in flocks, closely together, somewhat like red-heads and blue-bills, though rather more inclined to dart about irregularly. When wounded, they are exceedingly difficult to capture, being very expert divers and possessing extraordinary vitality. Their flesh is tender and well flavored. They remain with us quite late in the spring. No. 6 Chicago shot is about the proper sized shot to use for killing them.

CHAPTER XXIX.

Local name, "Butter-ball."

ADULT MALE : Bill much shorter than the head, comparatively narrow, deeper than broad at the base, gradually depressed toward the end, which is rounded.

Head rather large, compressed. Eyes of moderate size. Neck short and thick. Body compact, depressed. Feet very short, placed far back ; tarsus very short, compressed.

Plumage dense, soft, and blended. Feathers on the fore part of the head very small and rounded ; on the upper and hind parts linear and elongated, as they also are on the lateral and hind parts of the upper neck, so that, when raised, they give · the head an extremely tumid appearance, which is the more marked that the feathers of the neck immediately beneath are short. Wings very small, decurved, pointed. Tail short, graduated, of sixteen feathers.

Bill light grayish-blue. Iris hazel. Feet very pale flesh-color. Claws brownish-black. Fore part of the head of a deep rich green; upper part rich bluish-purple, of which color also are the elongated feathers on the fore part and sides of the neck; the hind part of the latter deep green; a broad band of pure white from one cheek to the other over the occiput. The colored parts of the head and neck are splendent and changeable. The rest of the neck, the lower parts, the outer scapulars, and a large patch on the wing, including the greater part of the smaller coverts and some of the secondary coverts and quills, pure white, the scapulars narrowly margined with black, as are the inner lateral feathers. The feathers on the anterior edge of the wing are black, narrowly edged with white. Alula, primary coverts, and primary quills deep black. The feathers on the rump gradually fade into grayish-white, and those of the tail are brownish-gray, with the edges paler and the shafts dusky.

Length to end of tail, 14½ inches; extent of wings, 23; weight, 1 pound.

ADULT FEMALE: The female is much smaller. The plumage of the head is not elongated, as in the male, but there is a ridge of longish fea-

thers down the occiput and nape. Bill darker
than that of the male. Feet grayish-blue, with
the webs dusky. Head, upper part of neck, hind
neck, back, and wings grayish-brown. A short
transverse white band from beneath the eye, and
a slight speck of the same on the lower eyelid.
Six of the secondary quills white, on the outer
web. Lower parts white, shaded into light grayish-
brown on the sides. Tail dull grayish-brown.

Length to end of tail, 13 inches; extent of
wings, 22¼; weight, 8 ounces.

Individuals of both sexes differ much in size
and in the tints of their plumage.

This pretty little species is common to nearly
every quarter of the United States, and frequents
both salt and fresh water. Local names : butter-
box, butter-ball, and little whistler. Their principal
food is fish, snails, etc.; consequently, their flesh is
never so well flavored as that of the vegetable-eating
ducks. They are rarely found in poor condition.
Buffle-heads are exceedingly quick-motioned in all
their actions, very expert in diving, which they
practise constantly when on the water, and fly
very swiftly, the action of their wings often making a whistling noise as they pass through the air.
They do not set their wings back and stop their

headway before alighting, as do most wild-fowl, but plump down, splashing the water on all sides, and, when the water is smooth, often slide along on its surface a considerable distance. They usually fly close to the water, and avoid crossing the land as much as possible. Being so little hunted, they are seldom very wary, and are not often found in very large numbers. Their note is a short, guttural quack. They are exceedingly retentive of life, and require hard hitting to secure them. The Western market-hunter seldom shoots at them, even when they come into his decoys, holding them in a sort of contempt, and considering them as rather insignificant game, though they find ready sale in market.

CHAPTER XXX.

In concluding these remarks on duck-shooting I must impress upon the reader's mind one final caution, without heeding which all his painstaking and skill in killing his game may prove useless. It is this: *Beware the crow!*

Crows are objects of inveterate enmity amongst all duck-shooters, and few ever venture within the circumscribed, and to them well-known, limits of shot-gun range without being forcibly reminded of the fact. The truth is, crows have a decided liking to the flavor of raw duck-flesh, and never hesitate to gratify their tastes whenever a favorable chance is offered. Being perhaps rather epicurean in their habits, they are so particularly partial to the heart and other tidbits that when they can get enough of these, all other portions are discarded as not worth eating. As a consequence, so many ducks are required to furnish even a small party of these fastidious yet thankless gluttons with a satisfactory repast, that hunters very naturally feel disinclined to cater for them when by any means it can be avoided.

On one occasion they destroyed for myself and companion over forty mallard ducks which we had killed one evening and left to be gathered the next morning, when the ice should be stronger, it being at the time we stopped shooting too weak to bear our weight, and yet too strong to be easily broken by our boat. We remained until dark, and, though we were back next morning but a few minutes after sunrise, the crows had arrived before us, and were then at their feast. Out of sixty odd ducks which we counted upon gathering, only fifteen were left us fit to be taken away; the rest were utterly spoiled. Two or three had been eaten by the minks and owls, as we had expected, and could see by their tracks upon the light snow which covered the ice; but to the crows we were chiefly indebted for our disappointment, and since that time opportunities for cancelling the debt have rarely been neglected. Though 1 have at various times, before and since, had from one to eight ducks destroyed by crows, such wholesale destruction as the above 1 have seen but once; and, learning caution from the experience, 1 mean never to see it again.

When you can collect your game in a pile,
you may keep it safe from their approaches by
leaving upon it your coat, vest, hat, or other object
which will produce in their minds suspicions of
a trap or other danger. At all events, if after a
hard day's work you cannot at the close of shooting conveniently carry away your game, remember
my parting word—*Beware the Crow!* and, either
by hiding the booty or scaring off the thief, secure the results of your labor.

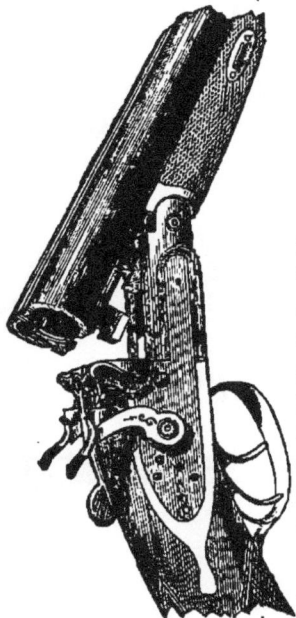

FIELD, COVER, AND TRAP SHOOTING,

By CAPTAIN A. H. BOGARDUS,

Champion Wing Shot of America.

A compendium of many years of experience, giving hints for skilled marksmen and instructions for young sportsmen, describing the haunts and habits of game birds, flight and resorts of water fowl, breeding and breaking of dogs, and everything of interest to the sportsman. The author knows a gun as Hiram Woodruff knew a horse. And he has the same careful and competent editor who put Woodruff's *Trotting Horse of America* into shape—Chas. J. Foster, so many years sporting editor of Wilkes' *Spirit of the Times.*

12mo, cloth, with steel portrait of the author, and an engraving of the Champion Medal, $2.00.

———o———

*** *To be had through any Bookseller; or will be mailed, post-paid, on receipt of price, by the Publishers,*

J. B. FORD & CO.,
27 Park Place, New York.

A WORK THAT EVERY SPORTSMAN WANTS.

Fur, Fin and Feather,

A QUARTERLY PERIODICAL,

DEVOTED TO

FIELD SPORTS

AND ANGLING,

BEING

A COMPILATION OF THE GAME LAWS

Of all the States of the Union and the Provinces of Canada,

WITH

INTERESTING ARTICLES ON HUNTING AND FISHING,

VALUABLE NOTES ON GAME AND FISH,

Descriptive Sketches of the Game Fields and Angling Waters of America,

And other Useful Information for Gunners and Anglers.

TERMS.

Single copy, one year, $2; six copies, $10; ten copies, $15; and for each additional copy at $2, the agent may retain 50 cents.

Single copies may be had of all gunsmiths, fishing-tackle dealers, or news agents, or will be mailed on receipt of price (50 cents) by

CHARLES SUYDAM, Publisher,

61 Warren St., New York City.

ORIENTAL POWDER MILLS,

MANUFACTURERS OF

SPORTING & BLASTING

GUNPOWDER,

No. 13 Broad Street,

BOSTON.

A. WILLIAMS, *Treasurer.*   C. O. FOSTER, *President.*

———◆———

DIAMOND GRAIN, in 1 lb. canisters.

FALCON DUCKING, coarse grain, clean and strong, for Breech-Loaders, in 75 lb. kegs and 6¼ lb. cans.

FALCON SPORTING, in 6¼ lb. cans.

DUCKING, for Sea Shooting, in kegs, half kegs, quarter kegs, 5 lb. cans, and 1 lb. cans.

WESTERN SPORTING, a superior Rifle Powder, in kegs, half kegs, quarter kegs, 5 lb. cans, and 1 lb cans.

———o———

Mining and Blasting Powder of superior quality, adapted to all kinds of work.

———◆———

Our Powder may also be obtained of our Agents :

P. PERLEY, 327 North Second Street, ST. LOUIS.

C. D. AUSTIN, 9 State Street, CHICAGO.

D. L. RANLETT & CO., 24, 26 and 28 Peter Street, NEW ORLEANS.

E. K. TRYON, JR. & CO., 19 North Sixth Street, PHILADELPHIA.

MITCHELL & WHITELAW, 70 Walnut Street, CINCINNATI.

JOHN E. LONG & CO.,   DETROIT.

———o———

SOLD BY ALL FIRST CLASS GUN DEALERS.

www.ingramcontent.com/pod-product-compliance
Lightning Source LLC
Chambersburg PA
CBHW021509210326
41599CB00012B/1192